Drought Early Warning and Forecasting

Drought Early Warning and Forecasting

Theory and Practice

CHRIS FUNK, PhD

United States Geological Survey Earth Resources Observation and Science Center (USGS EROS), Sioux Falls, South Dakota, U.S. & Climate Hazards Center, Department of Geography, University of California, Santa Barbara, California, U.S.

SHRADDHANAND SHUKLA, PhD

Climate Hazards Center, Department of Geography, University of California, Santa Barbara, California, U.S.

ELSEVIER

Elsevier
Radarweg 29, PO Box 211, 1000 AE Amsterdam, Netherlands
The Boulevard, Langford Lane, Kidlington, Oxford OX5 1GB, United Kingdom
50 Hampshire Street, 5th Floor, Cambridge, MA 02139, United States

British Library Cataloguing-in-Publication Data
A catalogue record for this book is available from the British Library

Library of Congress Cataloging-in-Publication Data
A catalog record for this book is available from the Library of Congress

ISBN: 978-0-12-814011-6

For Information on all Elsevier publications
visit our website at https://www.elsevier.com/books-and-journals

Publisher: Candice Janco
Acquisitions Editor: Louisa Munro
Editorial Project Manager: Michelle W. Fisher
Production Project Manager: Kumar Anbazhagan
Cover Designer: Christian J. Bilbow

Typeset by MPS Limited, Chennai, India

Dedication

We would like thank Sabina, Leah, and Mithilesh for their ongoing support and patience. Jim Verdin's decades of commitment to improved drought early warning provided critical inspiration for this work.

Contents

Foreword

Drought is among the most damaging, and least understood, of all "natural" hazards. Although recent drought-related disasters from Somalia, to Cape Town, to California, have contributed to a sense of urgency, droughts have not received commensurate attention within the hazards research and practitioner communities, unlike events such as hurricanes, floods, and earthquakes, which have direct and immediately visible impacts. Most countries, regions, and communities currently manage drought risk through reactive, crisis-driven approaches. Drought thus remains a "hidden" hazard and yet can span timescales from a few months and seasons to decades, and spatial scales from a few square kilometers to entire regions, with billions of dollars and thousands of lives and livelihoods affected or lost. Despite progress, there are still important limitations to our understanding and ability to predict various aspects of drought, including onset, duration, severity, recovery, and recurrence.

As has been noted, much is assumed about the vulnerability and capacity of those affected by environmental hazards. In the context of a changing climate, much is also assumed about the reliability of projections and, as importantly, about our understanding of present-day variability and how features such as ENSO might be changing. The assumptions that are being made by both researchers and practitioners warrant a more careful and updated explication of the state of knowledge, uncertainties, challenges, and benefits. This volume is a much needed and timely complement to the long-standing challenge of understanding the socio-ecological conditioning factors of drought-related risk, framed under the action-oriented concept of "early warning." As elucidated in the text, faster rates of environmental change, including compounding effects of heat stress/evaporative demand, and the increasingly complex development and ecological pathways through which drought impacts filter, may drive surprises and rapid transitions in which early warnings of emerging thresholds are increasingly critical. Improving predictions of the full life cycle of droughts requires a better understanding of how water, vegetation, and energy signals propagate through the ocean–atmosphere–land system, shedding light on the predictability of the various physical facets of drought, including precipitation, temperature, soil moisture, snow, and runoff.

As long recognized, and fully explored in this volume, a forecast by itself is not an early warning system. In a proactive framing an early warning system involves much more than development and dissemination of a forecast; it includes the systematic collection and analysis of relevant information about, and coming from, areas of impending risk that (1) informs the development of strategic responses to anticipate crises and crisis evolution; (2) provides capabilities for generating problem-specific risk assessments and scenarios; and (3) effectively works with and communicates options to critical actors for the purposes of decision-making, preparedness, and mitigation. Successful drought information systems have multiple subsystems supported by research in integrated risk assessment, communication, and decision support, of which early warning is a component and output. This volume brings to bear the experience of the authors, who have been engaged in early warning system development since the East African droughts of the mid-1980s, and systematically scrutinizes the "pros" and "cons" of existing and proposed systems, distilling lessons from past practices, landmark drought events, and advances in the field.

The authors capture key aspects of early warning design, including the importance of convergence of evidence, placing multiple indicators within consistent triggering frameworks, and the confounding factors of population, technology, and environmental change. Sources of physical information as noted in the text are derived from satellite data, in situ observations, land surface simulation, and model-based forecast skill. Key to the productive use of such information within early warnings is clearer delineation of sources of uncertainty, their reliability at different times of the year, and the integration of diverse environmental data sets into coherent databases such as the CHIRPS.

As the authors note, the volume aims at pragmatic goals—to provide a readable, accessible resource that is useful in both a classroom and in national, meteorological, and hydrological agencies. Several studies have identified the characteristics of predecisional practices that lead to effective communication over the long term. Key among these characteristics is the need to bring the delivery persons (e.g., extension personnel within local communities and the research community) to an understanding of what has to be done to translate current information, reliably, into local contexts, as well as the need to develop, support, and train a cadre of professionals who view the role of linking science, policy, and practices as a core goal over the long term. The chapters on practice on actionable information and decision-making networks provide an excellent grounding for the training of such professionals.

The authors are abundantly clear that the chapters do not presume to cover all aspects of drought risk and resilience management. They choose a more modest goal, highlighting the science and observations that underpin actionable impacts assessments and early warning. The depth to which these issues are pursued in the volume, while maintaining readability and accessibility, is impressive and very much welcomed. This book is an important contribution to the challenge of drawing information along the weather–climate continuum, from internal atmospheric variability to modes of climate. The chapters illustrate the need to not only understand and design information systems "for" change but also to design robust science-based systems that help us navigate "through" change. More than useful or even usable information is needed; what is required are pathways to improved decisions that thread through disaster risk reduction, adaptation, sustainability. Drought information systems along this continuum are investments rather than "costs." Such an informed drought early warning system would not be simply translational but, by design, transformative. This book provides robust and needed guidance on building such a path. The chapters in this volume will be vital tools in the quiver of effective early warning practitioners and researchers and for providing a space in which these communities can work together.

Roger S. Pulwarty
NOAA Physical Sciences Laboratory, Boulder, CO, U.S.

Preface

We are pleased to share with you *Drought Early Warning and Forecasting: Theory and Practice*. This book is intended to assist both students and drought early warning practitioners in understanding, developing, and applying drought early warning systems. Every year droughts impact millions of people and cause losses totaling billions of dollars. As our population and economies expand, so grows our exposure to drought risk. Increasing water demands and the impacts of climate change keep accelerating the need for effective drought management and drought early warning. Focusing on this second challenge, this manuscript gathers together many of the different components of drought early warning systems. The introductory chapters (Chapters 1–3) describe the historic nature of droughts while introducing drought early warning and early warning systems. Droughts are complex slow-onset multiscale disasters that impact many different sectors, and drought early warning systems must be correspondingly sophisticated. To this end, the next three chapters (Chapters 4–6) describe key "tools of the trade" (weather and climate forecasts, land surface models, and maps describing exposure and vulnerability). We then describe (Chapters 7–8) two theoretical frameworks central to understanding and monitoring droughts: atmospheric water demand and drought indices. The remaining "practice" chapters (Chapters 9–12) of the book address various aspects related to developing integrated systems. While some of this material has been drawn from our many years of experience with the Famine Early Warning Systems Network, the technical descriptions, general strategies, and lessons learned should be applicable in many drought early warning settings.

Drought early warning practitioners have a unique opportunity to use their skills to benefit society, guarding the lives and livelihoods of hundreds of thousands or even millions of people. These skills will be put to the test in the coming century, and we hope that this accessible "one-stop-shop" discussion of drought early warning science will prove valuable.

Acknowledgments

This work was supported by the U.S. Geological Survey Drivers of Drought and the Famine Early Warning Systems Network programs, U.S. Geological Survey (USGS) cooperative agreement #G14AC00042, United States Agency for International Development (USAID) cooperative agreement #72DFFP19CA00001, and National Aeronautics and Space Administration (NASA) grant #NNX16AM02G. USGS efforts were supported by a Participating Agency Program Agreement with USAID/FFP. Dr. Shukla also acknowledges National Oceanic and Atmospheric Administration (NOAA) Regional Integrated Sciences and Assessments (RISA)'s support through the California–Nevada Applications Program (Grant number: NA17OAR4310284).

The keen editorial assistance of Juliet Way-Henthorne, the Climate Hazards Center's Science writer, was absolutely vital to the completion and improvement of this book. She has tirelessly corrected our many mistakes and helped us meld these individual chapters into a cohesive whole. We could not have finished this book without her unstinting support and technical excellence. Sandra Cooper, the U.S. Geological Survey Bureau Approving Official, provided extremely valuable suggestions that greatly improved the overall manuscript. We are also indebted to Roger Pulwarty for his thoughtful foreword, which helps situate this book in the broader drought risk management landscape. Laura Harrison kindly provided material for Chapter 6, Tools of the Trade 4—Mapping Exposure and Vulnerability. We would also like to thank the many members or associates of the Climate Hazards Center, who have provided inspiration or chapter reviews, particularly, Pete Peterson, Tamuka Magadzire, Jim Verdin, Amy McNally, Rachel Green, Laura Harrison, and Natasha Krell all kindly provided internal reviews of selected chapters. Any use of trade, firm, or product names is for descriptive purposes only and does not imply endorsement by the U.S. Government.

CHAPTER 1

Droughts, governance, disasters, and response systems

History is something that very few people have been doing while everyone else was plowing fields and carrying water buckets.

Sapiens, Yuval Harari, p. 101

History for us waits silently in the basement of the National Museum of Ethiopia, in Addis Ababa (Fig. 1.1). There, Lucy, aka AL288-1, rests, as she has rested for some 3.2 million years. A member of *Australopithecus afarensis*, diminutive Lucy walked across the hot, low Afar Triangle in northeastern Ethiopia during the Pliocene Epoch, or more specifically, during the mid-Pliocene warm period. The Earth at this time was extremely warm, with carbon dioxide levels on par with those today (Raymo et al., 1996). High-resolution pollen data from Hadar, Ethiopia,

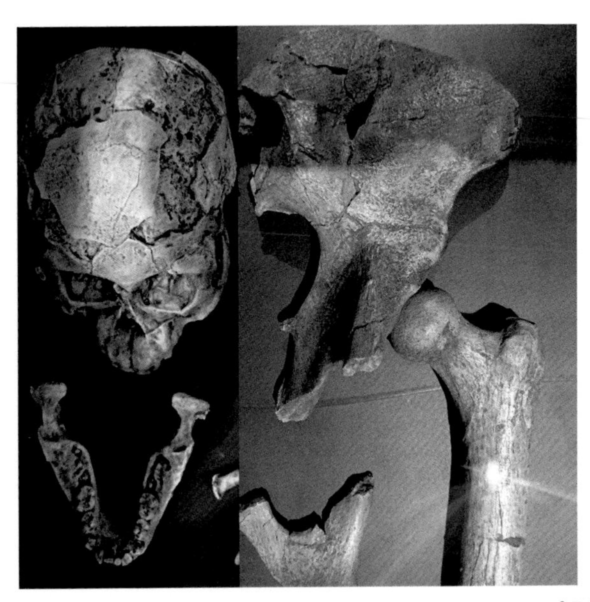

Figure 1.1 Skull and pelvis of Lucy, taken at the National Museum of Ethiopia, Addis Ababa. *Chris Funk.*

Drought Early Warning and Forecasting
DOI: https://doi.org/10.1016/B978-0-12-814011-6.00001-4

where Lucy was found, shows a large biome shift, with up to 5°C of cooling and a large increase in annual rainfall occurring between 3.4 and 2.9 million years ago (Bonnefille et al., 2004). This climate shift may have helped stimulate the evolution of bipedal proto-humans. Lucy's bones represent a species in transition. The shape of her knees and pelvis indicate bipedal locomotion, and the length of her arms was relatively short and her legs relatively long, compared to chimpanzees. Her brain, however, was quite small.

There are numerous theories as to why humans evolved bipedal motion. Bipedal locomotion is more efficient, and shifting forests and fragmented landscapes may have rewarded this innovation. Bipedalism may have made it easier to give birth to big-brained babies (Falk et al., 2012). Bipedal creatures have their hands free, enabling them to better produce and manipulate tools. This, in turn, may have led to a more protein-rich diet, potentially leading to an increase in brain size (Johanson and Edgar, 1996).

By 2.5 million years ago, our human ancestors were using simple tools. Our brains had grown prodigiously, necessitating the early birth of children and thus longer periods of childhood dependency. This may have necessitated strong social networks and language development. A single mother would have had a lot of trouble foraging for food and taking care of little children. The early birth of children, in turn, may have created unique opportunities for education, innovation, and cultural evolution. Tool use and communication may have helped move humans into a top predator position by 400,000 years ago. It was then that some human bands were hunting large game on a regular basis. By 300,000 years ago, we were using fire on a regular basis. Fire killed germs and parasites and made food easier to chew and digest, allowing our ancestors to eat a wider variety of food. Easier digestion may have led to shorter intestinal tracks in Sapiens and Neanderthals (Gibbons, 2007), reducing energy consumption, helping us fuel our massive brains. These huge brains, combined with our social proclivities, helped lead to evolutionary success, and out-migration from Africa about 70,000 years ago (Harari, 2015).

Humanity moved out of Africa (Fig. 1.2) and spread across Asia and Europe, with *Homo sapiens* supplanting Neanderthals, perhaps due to their advantageous communication skills and societal coordination. These advantages helped trigger the cognitive revolution, which allowed us to form large communities and learn, teach, and engage in complex behaviors. Slowly, these coordination skills grew, until early humans were able

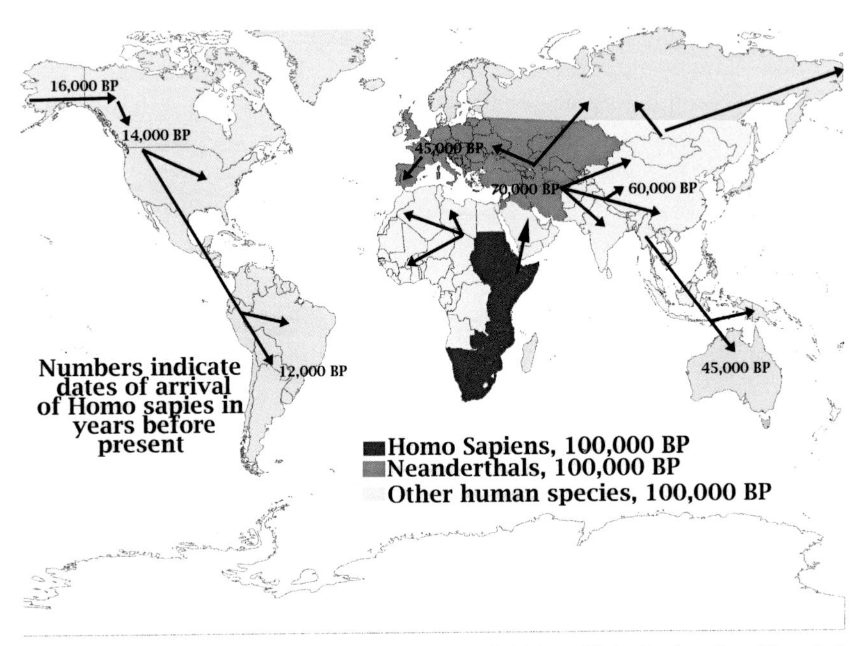

Figure 1.2 Expansion of *Homo sapiens* out of Africa. *Chris Funk, after Map 1 in Sapiens A Brief History of Humankind, Yuval Noah Harari.*

Figure 1.3 Sophisticated hunter–gatherers built a temple before agriculture. *Wikimedia.*

to assemble complex temples like Göbekli Tepe, which is in current-day southern Turkey near the Syrian border (Fig. 1.3). Built around 10,000 BCE, this Neolithic (Stone Age) temple predated agriculture. Immense 6-m, 20-ton T-shaped stone megaliths were fitted into sockets hewn in the bedrock to form circles. Intricate carvings of bulls, foxes, cranes, and people adorn pillars and totems. Göbekli Tepe supports the early importance of religion and social structure in human development. Before agriculture, pottery, writing, metallurgy, or the invention of the wheel, Neolithic hunter and gatherers built complex societies and buildings.

Soon, however, agriculture would change the world (Fig. 1.4). In the Middle East's Fertile Crescent, wheat and goats were domesticated by about 9000 BCE. c.7000 BCE, independent agriculture innovations led to the domestication of millet in current-day China. By 4000 BCE early varieties of maize, beans, and squash were being raised in Central America. In sub-Saharan Africa, agriculture emerged in the Ethiopian highlands, West Africa, and the Sahel (Diamond, 1997). Taro, bananas, and sugarcane were domesticated in New Guinea while South Americans began to harvest potatoes and manioc. In the southeastern United States, Native Americans raised sunflowers, sumpweed, and goosefoot.

With the rise of agriculture came an increased societal sensitivity to drought. With high mobility and low population densities, earlier hunter–gatherer populations were probably relatively resilient to most

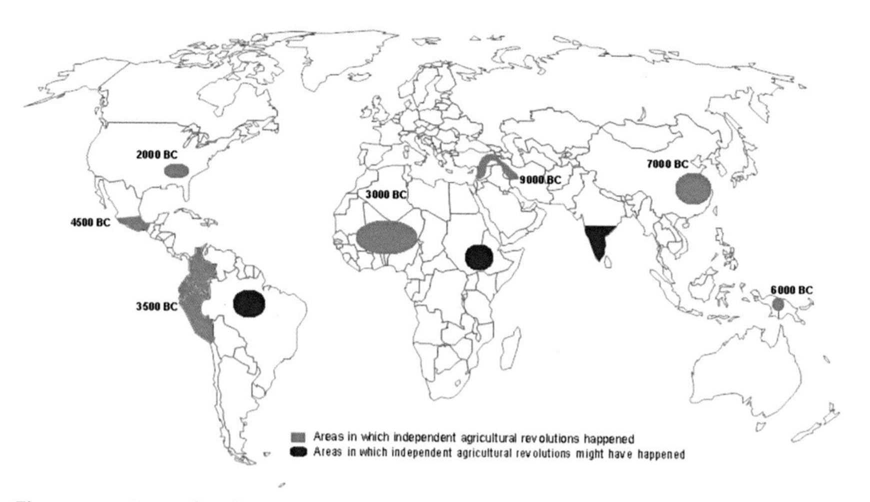

Figure 1.4 Sites of early agricultural development. *Chris Funk, after map in Sapiens A Brief History of Humankind, Yuval Noah Harari.*

climatic extremes. They could move toward more abundant rains, unlike the farmers of more recent times. The spatial compass of early agriculturalists extended to the few miles around their habitations. Villages and cities lived or died based on the strength of their harvests. For example, a study of drought stress variability in ancient Near Eastern agricultural systems finds "The emergence and decline of early civilizations is intrinsically tied to agricultural surplus production, either enabling a focus on technological progress and the accumulation of wealth or, in the case of insufficient yield, leading to hunger, violence, and war (Riehl et al., 2014)." One of the most severe crises in ancient times was the 4220 BP (2200 BCE) Holocene climate event, which was associated with severe aridity in North Africa, Egypt, the Middle East, the Indian Continent, and North America. This aridification event may have helped trigger the demise of the Old Kingdom in Egypt, the Indus Valley Civilization in India, the Akkadian Empire in Mesopotamia, and the Liangzhu culture in present-day China.

The most successful early civilizations supported dense populations by enabling them to devote resources to religious, administrative, military, artistic, and industrial activities. Droughts posed an existential threat to these societies. In response, various water and agricultural management practices were developed. In ancient Assyria, granaries and artificial irrigation helped overcome food shortages (Sołtysiak, 2016). In ancient Mesopotamia, where Hammurabi established his famous code in 1754 BCE, there developed a system of communal canals and irrigation works and a legal framework to govern these works (Kornfeld, 2009). In ancient Egypt, pharaohs effectively controlled equitable water distribution systems (primarily driven by manpower) that promoted social stability while providing supplies to cities and towns (Driaux, 2016). In ancient India, "deficiencies of rainfall were overcome by means of one or the other form of irrigation—rivers, canals, lakes, tanks, wells, artificial reservoirs, ponds and pools" (Date, 2008).

Given the acknowledged water dependencies of ancient agricultural societies, it comes as no surprise that they also showed a keen interest in drought early warning. Early agrarian populations grew rapidly and could only be sustained with extensive effort and stable access to water. In historic times, agricultural lands made up a small fraction of the Earth's land surface, 2% in CE 1400 (Marks, 2006). These lands needed to be defended, and when drought struck, there was little mobility, and the poor must make do.

As described by Yuval Harari:

From the very advent of agriculture, worries about the future became major players in the theatre of the human mind. Where farmers depended on rains to water their fields, the onset of the rainy season meant that each morning farmers gazed towards the horizon, sniffing the wind and straining their eyes. Is that a cloud? Would the rains come on time? Would there be enough? Would violate storms wash the seeds from the fields and batter down the seedlings? Meanwhile, in the valleys of the Euphrates, Indus, and Yellow Rivers, other peasants monitored, with no less trepidation, the heights of the water.

Sapiens, Yuval Harari, p. 101

Our agricultural foundations have guaranteed a strong dependence on adequate precipitation that remains with us today. In CE 1400 the Earth supported about 360 million people, whereas now it supports about 7.6 billion—a 21-fold increase. Today we farm about 12% of the Earth's ice-free surface (Ramankutty et al., 2008). In many parts of the world, technology has dramatically increased yields and altered farming practices. In others, poor farmers still sow and reap mostly by hand. In all of these regions, however, crops still require water. Technology can do little to overcome crop water deficits, and 80% of our croplands are still unirrigated. Water can be moved, but this is often expensive and energy intensive. In many parts of the developing world, furthermore, hydropower has emerged as a critical source of energy, a valuable but potentially fragile alternative to fossil fuels. Commodity markets, both local and global, can expose millions of poor people to price shocks, sometimes originating from droughts thousands of miles away. The poorest of these poor, often subsisting on 100 or 200 USD a year, typically spend more than half their incomes on food; when prices jump up, they must often do without food. In 2018 the city of Cape Town in South Africa faced "Day Zero," or the day when the city was expected to run out of water. Such crises are becoming increasingly common. In 2015 and 2016, Harare, Zimbabwe, grappling with the impact of an El Niño-related drought, faced similar conditions, as did Sao Paulo, Brazil. In 2018 Mexico's 21 million residents also faced shortages of running water.

In ancient times, settlements and cities grew along rivers and gave rise to great civilizations, and today, according to a recent World Bank report, we remain highly dependent on water availability (Damania et al., 2017). This report finds that "throughout much of the world, even moderate deviations from normal rainfall levels can cause large changes in crop yields." Rainfall shocks induce a cascade of effects that include lower agricultural production and deforestation. Ironically, in many areas, irrigation can result in the *paradox of supply*, when water is supplied too cheaply and consumed recklessly. Droughts can lead to severe undernutrition of children and mothers, producing childhood stunting, and an increase of low-birthweight children (Davenport et al., 2017; Grace et al., 2012, 2015), all of which can increase poverty and reinforce intergenerational health problems.

Droughts still contribute to severe hunger. In 2017 approximately 815 million people, roughly 1 out of every 10 people on our planet, were deemed by the United Nations Food and Agricultural Organization to be suffering from chronic hunger. Contemporaneous assessments of extremely food-insecure populations—those facing a real threat of famine without immediate emergency assistance—totaled 81 million, roughly one of every hundred human beings. Despite our great technological achievements, hunger and drought still plague the advance of humanity.

As populations expand and our climate becomes warmer and more variable, we will need improved drought early warning systems (DEWS). This book describes both the theory and practical methods required to create and effectively use these systems, helping to potentially save lives and livelihoods while mitigating some of the impacts of climate change. Over the next 20 years, we know that increasing population, economic growth, and rising air temperatures will increase demands for water, while rainfed and glacier-based water supplies are likely to become more erratic. We also know, however, that our ability to observe the Earth using satellites is rapidly increasing. With each passing year, we can better observe, explain, and predict weather and climate extremes. Richer social and computer networks are supporting enhanced communication and decision support. This information is being used successfully to help us anticipate, prepare for, and respond to 21st-century droughts. At the beginning of the 21st century, we are achieving what could only be dreamed of at the outset of the 20th century.

1.1 20th-century droughts—disasters and the El Niño–Southern Oscillation

The 20th century began with severe famines that killed tens of millions of people, helping to motivate the modern science of drought early warning. Between 1896 and 1902, in India and China, the monsoon rains failed, bringing destructive epidemics of malaria, bubonic plague, dysentery, smallpox, and cholera (Davis, 2002). Mortality estimates for India indicate 19 million people may have perished, while in China, it was estimated to be 10 million. The Bombay Government's "Report on the Famine in the Bombay Presidency" found that 1899–1900 harvests in the Bombay Deccan, Karnatak, and Gujarat provinces were only 4%–16% of normal. Indian authorities, held under rigid inflexible ideological British rule, failed to respond adequately to the extreme conditions. George Nathaniel Curzon, first Marquess Curzon of Kedleston, served as viceroy. In his zeal to suppress Home Rule, Curzon tightened press censorship, clamped down on education, and pitted Hindu against Muslim. For Curzon, financing the Boer War in South Africa was much more important than relieving the distress of the famine-stricken people of India. Writing at the time, and quoting data from the Lancet, William Digby wrote "This statement by what is probably the foremost medical journal in the world means that the loss of life thus recorded represented the 'disappearance' of fully one-half a population as large as that of the United Kingdom" (Davis, 2002).

For India the successive droughts of 1876–79, 1896–97, and 1899–1900 had a huge negative impact on productivity, livestock, and development. "Almost all the progress made in agricultural development since 1880 was nullified during the famines" (Davis, 2002).

India's terrible droughts, however, did help prompt major intellectual advances that eventually helped us understand, and sometimes anticipate, future climate extremes.

Gilbert Walker's appointment as Special Assistant to the Director General of the Indian Observatories in 1903 came as a surprise (Walker, 1997). Before being selected for the job, Gilbert "Boomerang" Walker had recently published an original and imaginative paper on these Australian spinning devices (boomerangs) in the Philosophical Transactions of the Royal Society (1897, 190, pp. 23–42). Walker had been selected because of his exceptionally strong mathematical capabilities. In 1904 he took charge of the India Meteorological Department. While organizing the various Indian weather observatories and services took up much of his time, he quickly turned to analyzing the accuracy of monsoon forecasts. Realizing that he could not predict the monsoon droughts analytically, he turned to the analysis of lagged correlations. Interestingly, this work led to fundamental advances in our understanding of one important feature of the mean global climate (the Walker Circulation), the most important quasiperiodic climate variation [the El Niño–Southern Oscillation (ENSO)], and a mathematical description of the expected autocorrelation structure of an autoregressive process (the Yule–Walker equations) (Katz, 2002).

Like previous meteorologists before him, Walker began the daunting prospect of looking for predictive anomalies that could be used to forecast Indian droughts. This led him to analyze variations in worldwide weather. Following the new work of the statistician Karl Pearson, he became a pioneer in the use of correlation in meteorology. By 1908 he was using multiple regression to predict monsoon rainfall. Examining the relatively rich set of global weather data available to a British meteorologist, Walker also looked for "Centers of Action" by examining extensive tables of autocorrelations and cross-correlations in sea-level pressure at multiple locations. From this analysis (Katz, 2002), he determined, "there is a swaying of pressure on a big scale backwards and forwards between the Pacific Ocean and the Indian Ocean, there are swayings, on a much smaller scale, between the Azores and Iceland, and between the areas of high and low pressure in the N. Pacific" (Walker, 1923). The swaying of pressure between the Pacific and Indian Oceans is now referred to as the "Southern Oscillation" (SO). The seesaw between the Azores and Iceland is known as the North Atlantic Oscillation, and the oscillation in the Pacific is referred to as the North Pacific Oscillation. All three of these patterns of climate variability have turned out to be critical drivers of climate and drought (Fig. 1.5).

Walker, however, pointed out that the influence of the SO seemed much greater and more persistent than the other two oscillatory patterns. Things have changed a lot since Walker flung his boomerangs and painfully worked out his cross-correlation tables by hand. Technology, the Internet, computers, and carefully compiled data sets now make it easy to analyze the importance of climate patterns like the SO. For example, with a few well-chosen clicks, we can compile a map of the correlation between local sea-level pressure values and the "SO Index"—an index

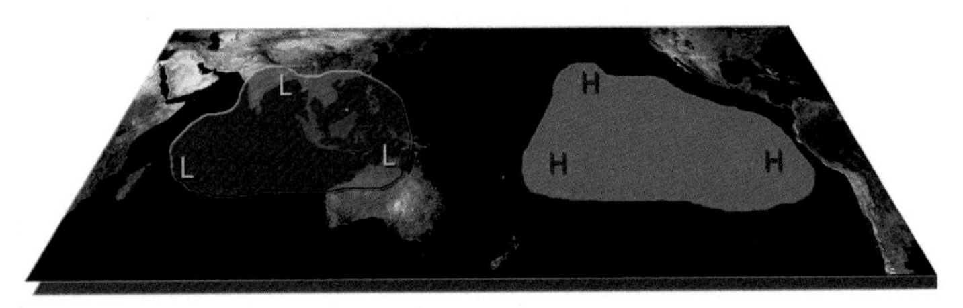

Figure 1.5 Centers of Action associated with the Southern Oscillation. *Chris Funk.*

measuring the difference in sea-level pressure between Tahiti in the South Pacific and Darwin, Australia. This pattern, first identified by Walker, identifies a quasi-global seesaw[1] in pressure between the west Pacific/Indian Ocean region and the east Pacific.

As owners of a vast international naval empire, the British were keen to track and analyze sea-level pressure data, like the data analyzed by Gilbert Walker, because it was strongly related to changes in surface winds. Understanding the mean (long-term average) structure of atmospheric circulations turns out to be a critical component of effective drought early warning. Understanding our "normal" climate provides an important foundation for understanding extremes. One key advance on this front was made by George Hadley (1685–1786). Hadley was an English lawyer and amateur meteorologist who proposed the atmospheric mechanism by which the trade winds are sustained. As a key factor in ensuring that European sailing vessels reached North American shores, understanding the trade winds was a matter of great importance at the time. Hadley was intrigued by the fact that winds, which should by all rights have blown straight toward the equator, had a pronounced westerly flow. Hadley (1735) began considering how the differential heating of the equator produced low atmospheric pressure, which drew in the low-level winds: "For let us suppose the Air in every Part to keep an equal Pace with the Earth in its diurnal Motion; in which case there will be no relative Motion of the Surface of the Earth and Air, and consequently no Wind; then by the Action of the Sun on the parts about the Equator, and the Rarefaction of the Air proceeding there from, let the Air be drawn thither from the N. and S. parts." Hadley then went on to introduce the idea of the conservation of angular momentum: "From which it follows, that the Air, as it moves from the Tropics towards the Equator, having a less Velocity than the Parts of the Earth it arrived at, will have a relative Motion contrary to that of the diurnal Motion of the Earth in those Parts, which being combined with the Motion towards the Equator, a N.E. wind be produced on this Side of the Equator, and S.E. on the other."

Hadley's theory turned out to be correct, and the structure he described is now known as the Hadley Circulation (Fig. 1.6). The Hadley Circulation described the North–South and up–down motions of the

[1] In writing the first draft of this sentence, we made a prescient typographic error—describing seesaw patterns. But it would be decades before scientists linked the atmospheric Southern Oscillation with the oceanic phenomenon known as El Niño.

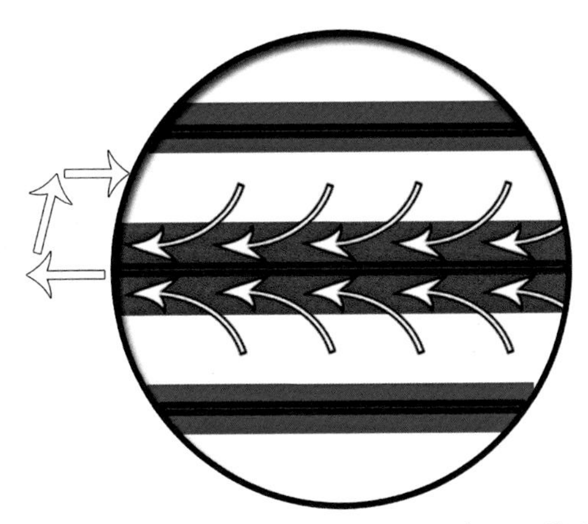

Figure 1.6 Schematic representation of the Hadley Circulation. *Chris Funk.*

atmosphere averaged across every longitude band. The differential heating of the equator combined with the conservation of angular momentum produces near-surface atmospheric convergence near the equator. This band of convergence is also associated with heavy rainfall and ascending atmospheric motions (red band in Fig. 1.6). Air parcels rise and move toward the poles (and the east). The conservation of angular motion again turns them into a strongly eastward wind field (the sub-Tropical Easterly Jet). These relatively warm parcels of air cool, radiating their extra warmth back out to space. This cooling makes the atmosphere very stable, and we find a tendency for air to subside (sink) in the subtropics, at latitudes of about 30-degree north and south. This latitude is where we tend to find most of the world's deserts and arid lands: the Sahara, Kalahari, Arabian Peninsula, Southwest and Central Asia, Atacama Desert, Australia, and the southwestern United States are associated with the sinking branch of the Hadley Circulation. Many severe droughts occur within or on the edges of these dry regimes, where conditions can slip from tentative to disastrous, sometimes for years on end.

Gilbert Walker's SO, however, focused on East–West variations between the eastern and western Pacific. Eventually, this would turn out to be a critical advance leading to effective drought forecasts. A key step toward that point, though, was describing the mean (long-term average) structure of the east–west atmospheric motions over the Pacific. This pioneering work was carried out by Jacob Bjerknes in the 1960s. He coined the term "Walker Circulation" and was the first person to describe the link between El Niño and the Southern Oscillation (Bjerknes, 1969). This coupled air-sea climate variation is now commonly referred to as the ENSO. Bjerknes started by noting that the eastern Pacific sea surface temperatures were exceptionally cold (Fig. 1.7), while the western Pacific/ Indian Ocean sea surface temperatures were much warmer. Because at the equator, we do not need to worry about the conservation of motion, Bjerknes theorized that this temperature gradient would set up a thermally direct circulation. Cold waters produce cold air with a relatively high density. Warm waters produce warm air with low densities. Winds at the equator will move toward the lower pressure. This wind pattern produces many positive feedbacks that help create the Indo-Pacific Warm Pool (Clement et al., 2005), while also reinforcing the strong east—west tropical temperature gradient between the Warm Pool and the equatorial east Pacific.

The "Warm Pool" is a term used to describe the world's warmest ocean waters in an area of the equatorial eastern Indian and Western Pacific Oceans (Fig. 1.8). In this rainiest region on the Earth, sea surface temperatures routinely exceed 29°C (84°F) and annual precipitation totals often exceed 3 m (129 in.) of rainfall. To the east of the Warm Pool the rapid persistent westward trade winds blow across the eastern Pacific. In the Pacific Ocean, these winds transport heat from the east Pacific into

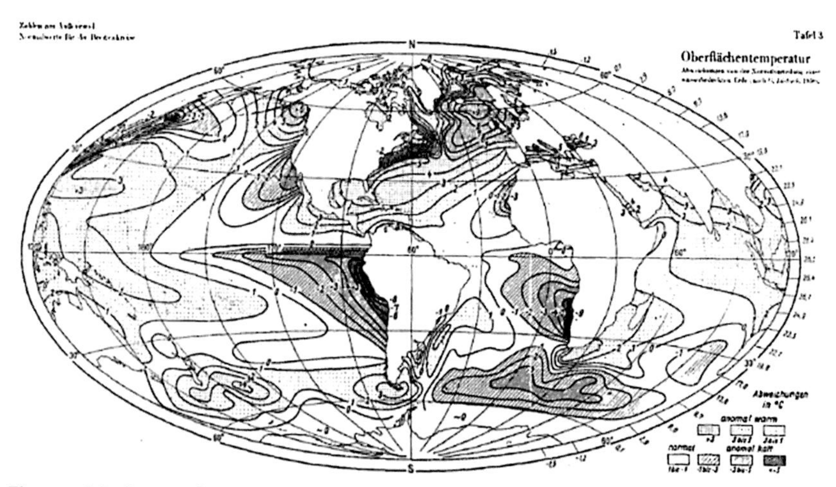

Figure 1.7 Sea surface temperature represented as deviation from the average at each latitude. *From Bjerknes, J., 1969. Atmospheric teleconnections from the equatorial Pacific. 97(3), 163–172—Figure 7.*

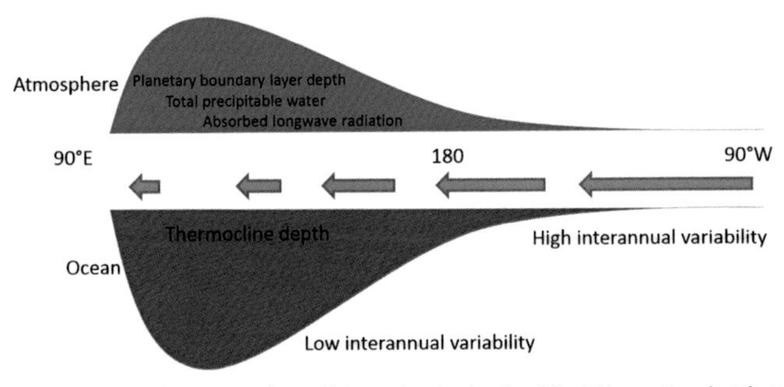

Figure 1.8 Schematic diagram describing the Indo-Pacific Warm Pool. *Chris Funk.*

the Warm Pool region. This acts to cool the east Pacific and warm the west Pacific. This warming pushes down the "thermocline" (the depth at which the ocean rapidly transitions to much cooler temperatures). This has the effect of producing very persistent warm ocean conditions. Above the Warm Pool, we find a similar convergence of warm, moist air. The buildup of this warm, moist air increases the depth of the planetary boundary layer (the lowest well-mixed layer of the atmosphere, which tends to be humid over the oceans). This increase in wet, warm air has several important feedbacks that help warm the Warm Pool and maintain the Walker Circulation. The thick boundary layer over the Warm Pool holds a lot of water (about $50\,\mathrm{kg\,m^{-2}}$), and this water vapor is a very active greenhouse gas. When there is abundant water vapor over the Warm Pool, this moisture acts as positive feedback, warming the Warm Pool by increasing the amount of longwave (infrared) radiation absorbed by the ocean.

A less direct but equally important feed involves the atmospheric response to the conditionally unstable warm moist air above the Warm Pool. Intense precipitation occurs easily under such conditions. Condensing water vapor releases a great deal of energy, and this heating produces rising air near the intersection of the Indian and Western Pacific Oceans (Fig. 1.9). This air rises and cools and then tends to move toward the eastern Pacific and western Indian Oceans, where it sinks producing a tendency for hot dry surface conditions. This subsidence produces high surface pressures, which results in a surface pressure gradient between the east Pacific and the west Pacific. This gradient supports the easterly trade winds, reinforcing heat and water vapor transports into the Warm Pool and out of the eastern equatorial region, completing the Walker

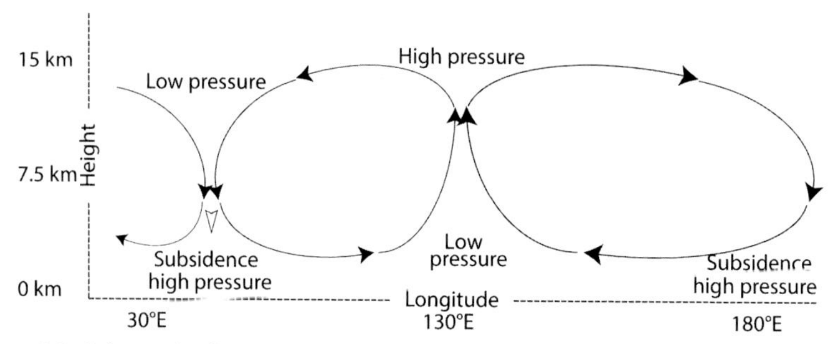

Figure 1.9 Schematic diagram overturning Walker Circulation. *Chris Funk.*

Circulation. This convergence of heat and water, in turn, feeds the Warm Pool's heavy precipitation.

Another subtle but critical feature of the equatorial ocean acts to enhance the climatological (long-term average) sea surface temperature gradient between the eastern and western Pacific (Fig. 1.7). This mechanism is known as the Bjerknes feedback. A critical feature of this feedback is how the equatorial ocean responds to easterly winds. These winds induce Ekman pumping, which draws up cool subsurface waters. This vertical mixing keeps the eastern Pacific cooler than the Warm Pool, reinforcing the Walker Circulation (Figs. 1.8 and 1.9

Sometimes, however, a westerly wind burst over the Pacific can trigger an "El Niño event." The term "El Niño" refers to the "Christ Child" and was associated with warm waters in the far eastern Pacific that used to appear in December, bringing bountiful harvests to fishermen in Chile and Peru (Glantz, 2001). These warm waters create a dramatic reversal of the equatorial trade winds over the east Pacific, shutting down the cooling produced by Ekman pumping and triggering a dramatic increase in moisture convergence and precipitation over the eastern Pacific. This oceanic El Niño, combined with the atmospheric SO identified by Gilbert Walker, results in a coupled ocean-atmospheric phenomenon—the ENSO.

An ENSO event produced the severe drought that led to India's turn of the century drought, setting the stage for famine (Davis, 2002). Another ENSO event in 1982–83 helped trigger a massive drought that stretched across the Sahel region of West Africa and into Sudan and Ethiopia, helping to fuel a massive outbreak of famine. Again in 1997/98 and 2015/16, large El Niños strongly modulated the Walker Circulation, helping to produce droughts and floods and increase air temperatures. In

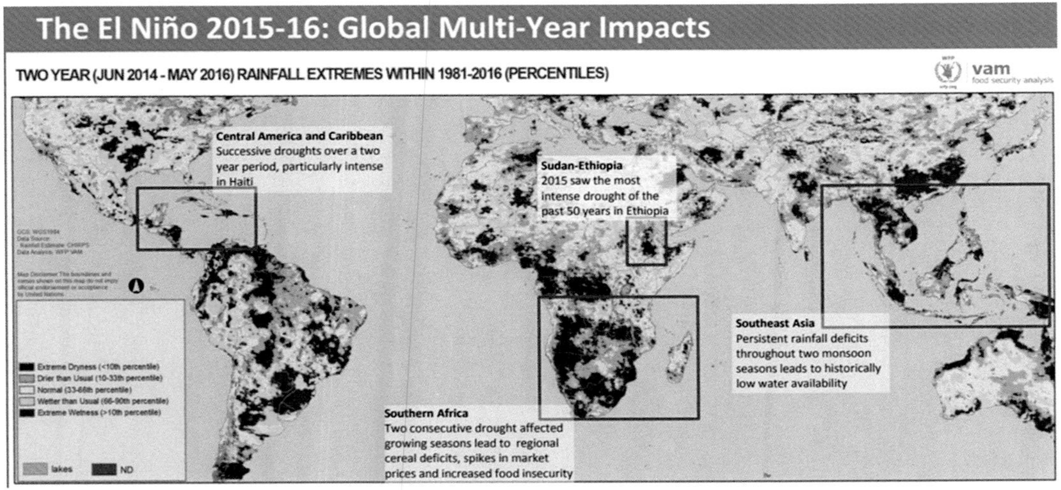

Figure 1.10 2015–16 Droughts associated with El Niño food security impacts. *Rogerio Bonifacio, World Food Programme.*

2015/16 food security experts from the World Food Programme linked El Niño (Fig. 1.10) with widespread dry conditions across Central and South America, southern Africa, India, Southeast Asia, and the Maritime Continent. In Ethiopia and southern Africa the extreme 2015/16 event drastically reduced runoff, crop production, and pasture conditions (Funk et al., 2016, 2017), leading to an enormous (~26 million person) increase in the level global acute food insecurity (Funk et al., 2018). Thailand suffered more than $500 million dollars in agricultural losses (Christidis et al., 2018). In India, in May of 2015, a heat wave led to the death of thousands (Di Liberto, 2015), as premonsoon temperatures soared above 110°F (43°C) over much of the country, and some regions saw temperatures as high as 117.7°F (47.6°C). The following monsoon season in India was very poor, negatively impacting more than 330 million people.

Between 1915, when "Boomerang Walker" pored over statistical tables looking for clues to understand and anticipate Indian Monsoon failures, and 2015 when the world climate modeling centers successfully predicted a severe El Niño associated with poor Indian Monsoonal rains, all the elements supporting effective drought early warning were put in place. Standing on the shoulders of giants, we have a much deeper understanding of how our global circulation systems work, and how they may be perturbed by climate variations like El Niños. Decades of efforts assembling observational systems based on weather stations and satellite Earth observations allow us to track global weather variations in near real time. Huge investments in climate modeling and computation now allow us to make skillful predictions days, weeks, and sometimes even months into the future.

1.2 21st-century droughts—developing effective early warning systems

Droughts are "slow-onset disasters." Unlike quickly striking crises, such as earthquakes, they creep up slowly through the gradual accumulation of water deficits. This creates a unique opportunity for effective early warning. At the same time, the early identification of droughts can be very challenging, in part because it can be unclear as to "when" a drought has really "happened." As a metaphor, consider the fable of the frog in a pot of water. According to the fable, but not experimental analysis, a frog placed in a slowly warming pot will sit placidly until expiration. Like the proverbial frog in the pot of warming water, droughts tend to sneak up on us slowly. A complicating factor (discussed more completely in the next chapter) is the fact that droughts can be defined in many ways. While always involving a shortage of available water, how this shortage arises and the measure of its severity depends on many factors, such as the intended use of the water, the vulnerability, coping capacity and exposure of the human-environmental system, and the historical and cultural context. In one context a 100 mm (4 in.) rainfall deficit might mean the difference between death and destruction or salvation and stability, but in another setting, say on the windward side of a tropical island, such a difference might not even be noticed.

As we shall see, successful 21st-century DEWS are effective, and interesting, because they successfully integrate and express information across many physical and social domains. This integration is both substantive and linguistic. For example, effective DEWS will combine (Fig. 1.11) information from multiple disciplines and data sources in ways that take advantage of and interpret the relevant context and setting. Take, for example, the "simple" example of a rancher in Kansas or a Masai pastoralist living in southern Kenya. A drought early warning specialist might be interested in information provided by the following disciplines: veterinary science, drylands ecology, herd dynamics, micro- and macroeconomics, hydrology, meteorology, climate modeling, climate dynamics, paleoclimate, and remote sensing. Typically, only a relatively modest understanding of these vast fields is required, but this information can be very important. Assume

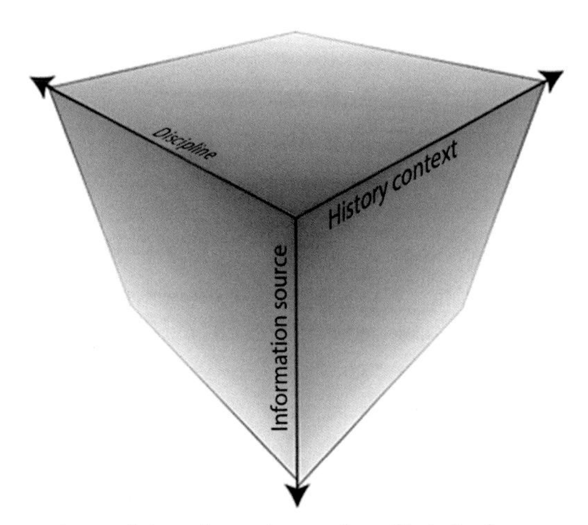

Figure 1.11 Dimensions of drought early warning. *Chris Funk.*

that we are interested in drought as it affects a small independent rancher in Kansas or an itinerant pastoral family in Africa. Effective drought early warning could be informed by our understanding of the relation between high temperatures, water stress, and limited grazing potential and cattle health. When do high temperatures start having a large negative impact on herds? Is water being provided via irrigation, rainfed streams and ponds, snowmelt from glaciers? Are the herds grazing rangeland grasses, or consuming large quantities of purchased grain? How do the ecosystems of interest respond to water stress? How might these water-related stresses be compounded by environmental conditions such as high temperatures and atmospheric evaporative demand or degraded pastures? How might social conditions such as large herd densities or a limited ability of the herder or rancher to respond to water stress amplify or minimize the impact of a given water shortage? How might the economic context modify the impact of a given climate shock? How poor are households being considered? What are their financial reserves? How sensitive is the given financial market to shocks in production? Do prices of herd animals and inputs such as fodder vary dramatically?

Connecting the dots across sets of questions like these is daunting, and such challenges give birth to DEWS. DEWS, like the U.S. National Integrated Drought System (NIDIS), which focuses on the Continental United States, or the U.S. Agency for International Development's food security–oriented Famine Early Warning Systems Network (FEWS NET), combine the abilities of many disciplinary experts, who combine information from many sources, in ways that take into account the history and context of a given application (see Chapter 3: Drought Early Warning Systems).

"System" used in this context is more similar to a hospital than a computer program or flowchart. When the ambulance pulls up to the hospital, one hopes to have on hand the combined expertise from specialists of many disciplines. These experts will not only combine information from an array of multiple data sources but also need to take into account the life history and context of the patient and their ailment.

Each step across a discipline or data source creates a potential challenge in communication. Effective translation provides one of the mainstays of effective drought early warning. In the 21st century, we can achieve so very much more than Gilbert Walker might have dreamed of. We can predict El Niños with a high degree of certainty. These predictions are only useful if we can predict changes in regional precipitation and temperatures. Using climate models and satellite-enhanced weather observing systems, we can predict and monitor precipitation and temperature with reasonable levels of accuracies in many places. The utility of this information is often limited, unless we can translate these weather fluctuations into impacts or shocks—like reductions in crop yields, or pasture biomass, or potential hydropower production. And the utility of such impact assessments is amplified by context-sensitive communication and interpretation. Knowing where a given reduction in yields/pasture biomass or potential hydropower production might have the greatest impact helps guide effective responses. Knowing how to package and communicate drought early warning information is a central component of effective DEWS, and this includes communication between domain experts and the stakeholders who use the warnings provided.

While the challenges involved in implementing effective DEWS are challenging, they are not insurmountable, and good DEWS inevitably evolve and improve over time. As our human population grows, economic activities increase, and on a rapidly warming planet, crop and water stresses will increase, and we will need better drought monitoring and prediction systems. This book describes some of the components of such systems, with a general focus on the Earth system science aspects of DEWS. Chapters 2–6 explore and define various pillars of successful DEWS. What are some of the basic challenges and opportunities associated with drought early warning (Chapter 2: Drought Early Warning—Definitions, Challenges, and Opportunities)? How do well-developed DEWS such as NIDIS or FEWS NET work (Chapter 3: Drought Early Warning Systems)? What are the primary tools of the trade (Chapters 4–6)? Chapter 7, Theory—Understanding Atmospheric Demand in a Warming World, and Chapter 8, Theory—Indices for Measuring Drought Severity, explore the scientific aspects of effective drought prediction. What causes droughts and how can we measure them? Chapters 9–11 then apply our understanding.

References

Bjerknes, J., 1969. Atmospheric teleconnections from the equatorial Pacific. J. Phys. Oceanogr. 97(3), 163–172.

Bonnefille, R., Potts, R., Chalié, F., Jolly, D., Peyron, O., 2004. High-resolution vegetation and climate change associated with Pliocene *Australopithecus afarensis*. Proc. Natl. Acad. Sci. U.S.A. 101, 12125–12129.

Christidis, N., Manomaiphiboon, K., Ciavarella, A., Stott, P.A., 2018. The hot and dry April of 2016 in Thailand. Bull. Am. Meteorol. Soc. 99, S128–S132.
Clement, A.C., Seager, R., Murtugudde, R., 2005. Why are there tropical warm pools? J. Clim. 18, 5294–5311.
Damania, R., et al., 2017. Uncharted Waters: The New Economics of Water Scarcity and Variability. World Bank.
Date, R., 2008. Water-management in ancient India. Bull. Deccan Coll. Res. Inst. 68/69, 377–382.
Davenport, F., Grace, K., Funk, C., Shukla, S., 2017. Child health outcomes in sub-Saharan Africa: a comparison of changes in climate and socio-economic factors. Global Environ. Change 46, 72–87.
Davis, M., 2002. Late Victorian Holocausts. Verso.
Diamond, J., 1997. Guns, Germs and Steel. W.W. Norton.
Di Liberto, T., 2015. India heat wave kills thousands. ClimateWatch Magazine June 9, 2015.
Driaux, D., 2016. Water supply of ancient Egyptian settlements: the role of the state. Overview of a relatively equitable scheme from the Old to New Kingdom (ca. 2543–1077 BC). Water Hist. 8, 43–58.
Falk, D., Zollikofer, C.P.E., Morimoto, N., Ponce de León, M.S., 2012. Metopic suture of Taung (*Australopithecus africanus*) and its implications for hominin brain evolution. Proc. Natl. Acad. Sci. 109.
Funk, C., et al., 2016. Assessing the contributions of local and east Pacific warming to the 2015 droughts in Ethiopia and southern Africa. Bull. Am. Meteorol. Soc. 97, S75–S80.
Funk, C., et al., 2017. Anthropogenic enhancement of moderate-to-strong El Niños likely contributed to drought and poor harvests in southern Africa during 2016. Bull. Am. Meteorol. Soc. 37.
Funk, C., et al., 2018. Examining the role of unusually warm Indo-Pacific sea surface temperatures in recent African droughts. Q. J. R. Meteorolog. Soc. 144, 360–383.
Gibbons, A., 2007. Food for thought (June, pg 1558, 2007). Science 317, 1036.
Glantz, M.H., 2001. Currents of Change: Impacts of El Niño and La Niña on Climate and Society. Cambridge University Press.
Grace, K., Davenport, F., Funk, C., 2015. Linking climate change and health outcomes: examining the relationship between temperature, rainfall and low birth weight in Africa. Global Environ. Change 35, 125–137.
Grace, K., Davenport, F., Funk, C., Mcnally, A.P., 2012. Child malnutrition and climate in sub-Saharan Africa: an analysis of recent trends in Kenya. Appl. Geogr. 35.
Hadley, G., 1735. On the cause of the general trade winds. Philos. Trans. R. Soc 34, 58–62.
Harari, Y., 2015. Sapiens. HarperCollins.
Johanson, D.C., Edgar, B., 1996. From Lucy to Language. Simon & Schuster.
Katz, R.W., 2002. Sir Gilbert Walker and a connection between El Niño and statistics. Stat. Sci. 17, 97–112.
Kornfeld, I.E., 2009. Mesopotamia: a history of water and law. The Evolution of the Law and Politics of Water. Springer, pp. 21–36.
Marks, R.B., 2006. The Origins of the Modern World: A Global and Ecological Narrative From the Fifteenth to the Twenty-First Century. Rowman & Littlefield Publishers, p. 220.
Ramankutty, N., Evan, A.T., Monfreda, C., Foley, J.A., 2008. Farming the planet: 1. Geographic distribution of global agricultural lands in the year 2000. Global Biogeochem. Cycles 22, n/a-n/a.

Raymo, M.E., Grant, B., Horowitz, M., Rau, G.H., 1996. Mid-Pliocene warmth: stronger greenhouse and stronger conveyor. Mar. Micropaleontol. 27, 313–326.
Riehl, S., Pustovoytov, K.E., Weippert, H., Klett, S., Hole, F., 2014. Drought stress variability in ancient Near Eastern agricultural systems evidenced by δ13C in barley grain. Proc. Natl. Acad. Sci. U.S.A. 111, 12348–12353.
Sołtysiak, A., 2016. Drought and the fall of Assyria: quite another story. Clim. Change 136, 389–394.
Walker, G.T., 1923. Correlation in seasonal variations of weather, VIII: A preliminary study of world weather. Memoirs of the Indian Meteorological Department. Indian Meteorological Department.
Walker, J., 1997. Pen portrait of Sir Gilbert Walker. Weather 52, 217–220.

CHAPTER 2

Drought early warning—definitions, challenges, and opportunities

2.1 Definitions—supply and demand, the many flavors of "dry"

Man, if you have to ask what jazz is, you'll never know.

Louis Armstrong.

Droughts are enigmatic (Wilhite, 1993), creeping, and hard to define despite the fact that they are one of the most widespread and damaging types of natural disasters (Wilhite and Glantz, 1985). As discussed in Chapter 1, Droughts, Governance, Disasters, and Response Systems, anticipating, monitoring, and responding to droughts has been a major challenge for humanity since the Neolithic era some 9000 years ago. The purpose of this book is to provide a comprehensive and integrative discussion of the physical science underlying the many components of an integrated drought early warning system (DEWS). Our focus is on providing a description of the technologies, principles, and application strategies that support successful 21st-century DEWS, such as those listed in Table 2.1. We hope to provide a single resource that can describe many of the key aspects of operational DEWS, and how these different constituents may or may not fit together. The reader should note that there are many excellent books that recount the historical development and more detailed scientific aspects of drought monitoring. Table 2.2 provides an incomplete list of some of these excellent resources. Integrated approaches to drought risk reduction stand on three pillars (Wilhite and Pulwarty, 2017): monitoring and early warning and information delivery systems, vulnerability and impact assessment, and mitigation and response. This book focuses on half of the first pillar: monitoring and early warning, with some discussion of how effective monitoring and early warning can successfully inform impact assessments and effective mitigation and response. As we will see, just a cursory description of all the components entering 21st-century monitoring and early warning systems draws on a wide range of subjects,

Drought Early Warning and Forecasting
DOI: https://doi.org/10.1016/B978-0-12-814011-6.00002-6

Table 2.1 Examples of drought early warning systems, including those listed in Pulwarty and Sivakumar (2014).

Early warning system or sites	Website
FAO GIEWS	http://www.fao.org/giews/en/
U.S. National Integrated Drought Information System	www.drought.gov
U.S. Drought Monitor	http://droughtmonitor.unl.edu/
California/Nevada Climate Applications Program	https://scripps.ucsd.edu/programs/cnap/
Famine Early Warning Systems Network	www.fews.net
Global Integrated Drought Monitoring and Prediction System	http://drought.eng.uci.edu/
South Asia Drought Monitor	https://sites.google.com/a/iitgn.ac.in/high_resolution_south_asia_drought_monitor/
Monitoring Agricultural Resources	https://ec.europa.eu/jrc/en/mars
IGAD Climate Prediction and Applications Centre	http://www.icpac.net/
World Food Programme	http://vam.wfp.org/sites/seasonal_monitor/
Permanent Interstate Committee for Drought control in the Sahel	http://www.agrhymet.ne/eng/
Southeast Asia Drought Monitoring System	http://dms.iwmi.org/
southern African Development Community, Directorate Climate Services Centre	http://www.sadc.int/sadc-secretariat/services-centres/climate-services-centre/
European Drought Observatory	http://climate-adapt.eea.europa.eu/metadata/portals/european-drought-observatory-edo
GEOGLAM Crop Monitor	https://cropmonitor.org/

FAO GIEWS, Food and Agricultural Organization Global Information and Early Warning System.
Source: Information compiled by Shukla and Funk (2019).

including at least meteorology, climatology, oceanography, satellite remote sensing, agronomy, hydrology, nutrition, health, economics, and food security. As described in a 2014 paper (Pulwarty and Sivakumar, 2014), early warning provides a critical component of modern DEWS. As described in this study, an early warning system is much more than just a forecast—"it is a linked risk information system and communication

Table 2.2 Relevant books describing droughts and its early warning systems.

Drought-related books	Date, authors, or editors
Drought: Past Problems and Future Scenarios	2012, Sheffield, J., Wood, E.F.
Remote Sensing of Drought: Innovative Monitoring Approaches	2012, Wardlow, B.D., Anderson, M.C., Verdin, J.P.
Drought and Water Crises: Integrating Science, Management, and Policy	2017, Wilhite, D., Pulwarty, R.S.
Drought: a Global Assessment	2000, Wilhite, D.
Drought and Water Crises: Science, Technology, and Management Issues	2005, Wilhite, D.
Drought: Research and Science-Policy Interfacing	2015, Paredes-Arquiola, J., Haro-Monteagudo, D., Van Lanen, H.
Managing the Risks of Extreme Events and Disasters to Advance Climate Change Adaptation: Special Report of the Intergovernmental Panel on Climate Change	2012, Field, C.B., et al. https://www.ipcc.ch/pdf/special-reports/srex/SREX_Full_Report.pdf
Famine Early Warning Systems and Remote Sensing Data	2008, Brown, M.

Source: Information compiled by Shukla and Funk (2019).

system that actively engages communities involved in preparedness." In this book, we stress that effective early warning of drought risk requires the integration of information across multiple disciplines, origins, and time scales, presented in a way that can be effectively used by stakeholders. Modern drought early warning begins with a clear but complex definition of drought and its impacts.

2.2 Droughts—when water demand exceeds water supply

Originally, the term drought comes from the old English term "drugað" and the Germanic root *dreug, meaning "dry." In modern times, we generally use this term to refer to a period of time, in a specific place, when there is not enough water. This latter phrase, however, is very vague, because it implies both variations in the overall water supply and the overall demand for water. While both supply and demand may have important social, economic, and sectoral permutations, it is worth starting with a generic "bucket model" describing the amount of water in a column of soil, watershed, or reservoir as a function of storage (*S*), supply described here as rainfall (R), and a rate of extraction or demand, described here using the rate of actual evapotranspiration. Between some time "i" and time "$i+1$," the change in storage will be equal to $R_i - \mathrm{ET}_i$, and we can describe our bucket model system using a simple equation: $S_{i+1} = S_i + R_i - \mathrm{ET}_i$. Like a bank account, S_i quantifies our reserves. If S_i is large compared to our supply and demand terms (R_i and ET_i), we will experience droughts infrequently. If S_i is small, the system being examined will be drought prone. Multidisciplinary drought early warning science focuses on understanding, modeling, and predicting the storage supply and demand terms that produce shortages of water. The opportunities associated with drought early warning are many: droughts are the most common natural disaster, and the incremental nature of water deficits ensures that droughts are "slow-onset disasters," typically providing some window for the early prediction and identification of extreme events.

Seminal definitions of drought (Wilhite and Glantz, 1985; Wilhite, 1993) tend to emphasize four key aspects: the slow onset of droughts, the multidisciplinary and multisectoral nature of droughts, the multidimensionality of droughts (intensity, duration, and extent), and the complexity of drought impacts. Updates to this work tend to emphasize the need to develop integrated decision support systems (Pulwarty and Sivakumar, 2014; Wilhite and Pulwarty, 2017) and the ability to develop early warning systems that effectively utilize both climate forecasts and Earth observations (Mariotti et al., 2013; Mo et al., 2012; Shukla et al., 2013). In this chapter, we describe the four key aspects of droughts, then briefly summarize some of the major challenges and opportunities facing 21st-century DEWS.

2.3 Slow-onset disasters

While complex, and difficult to define and identify, droughts are one of the most common natural disasters. In 2016, for example, a review of disaster statistics (Guha-Sapir et al., 2017) found that droughts accounted for 69.1% of all the people affected by disasters. Three hundred and ninety-three million people, the greatest number on record, were affected by droughts in 2016. Most of these people (330 million) lived in India, which experienced severe El Niño-induced rainfall deficits. Like most droughts, the 2016 crisis in India resulted in multiple days of dry weather. Unlike fires or floods, droughts are *slow-onset disasters*. This obviously not only provides an opportunity for early warning but also poses a barrier to action. When do we transition from a streak of hot, dry weather to disaster? As we shall see later, answering this question requires context. Droughts involve unmet water demands. Quantifying these demands and the impacts of water deficits is a critical component of modern drought early warning. Droughts are multifaceted, largely because the impacts of drought can arise in many ways, "being direct or indirect, either singular or cumulative, immediate or delayed" (Asfaw, 1983, in Wilhite, 1993). The multidimensional and multidisciplinary nature of droughts makes them difficult to understand and identify.

2.4 Quantifying drought magnitude in multiple dimensions

Drought *magnitudes* are typically categorized using three dimensions—their temporal *duration*, *severity*, and *extent* (Sheffield and Wood, 2012; Wilhite and Glantz, 1985), with timing being an important aspect as well. Drought impacts are typically defined in reference to a specific sector or discipline, such as agriculture. Examples of *duration*, *severity*, and *extent* types of magnitude categorization might include "Observers fear that two years of drought will soon become famine in Somalia"[1] (duration), or "Ethiopia struggles with the worst drought in 50 years"[2] (severity), or "More than half of Continental states experiencing extremely dry conditions"[3] (extent).

One widely used system in the United States is the Drought Classification used by the U.S. Drought Monitor[4] (Svoboda et al., 2002). This classification system ranges from not-dry to abnormally dry (D0), moderate drought (D1), severe drought (D2), extreme drought (D3), and exceptional drought (D4). These classifications are based on multiple indicators: the Palmer Drought Severity Index (Chapter 8: Theory—Indices for Measuring Drought Severity), Climate Prediction Center Soil Moisture Percentiles, U.S. Geological Survey streamflow percentiles, Standardized Precipitation Index (Chapter 8: Theory—Indices for Measuring Drought Severity) values, and objective drought indicator values. These quantitative metrics are augmented by local reports from more than 350 expert observers in a consultative process.

[1] https://news.nationalgeographic.com/2017/03/drought-somalia-puntland/

[2] https://www.telegraph.co.uk/news/2016/04/23/ethiopia-struggles-with-worst-drought-for-50-years-leaving-18-mi/

[3] https://www.huffingtonpost.com/2012/07/06/us-drought-2012-heat-wave_n_1654908.html

[4] http://droughtmonitor.unl.edu/AboutUSDM/DroughtClassification.aspx

Fig. 2.1 shows the April 10, 2018 U.S. Drought Monitor Map. The map is dominated by La Niña-like drought patterns affecting the southwestern and southeastern United States. The central southwest (Utah, Colorado, Oklahoma, Kansas, Texas, and New Mexico) comprises the driest states, with dry conditions ranging from moderate to exceptional drought.

The U.S. Drought Monitor data can be expressed as a time series. Fig. 2.2 (top) shows the fraction of the continental United States experiencing dry (D0), moderate drought (D1), severe drought (D2), extreme drought (D3), or exceptional drought (D4). This figure manages to convey both the extent and overall magnitude of the U.S. drought. The greatest overall extent of drought appears during 2012 and early 2013, when most of the United States experienced dry conditions (as discussed in more detail later). The greatest recent spatial extent of exceptional drought (D4) occurred during 2011 when severe dryness encompassed virtually all of Texas and much of its neighboring states. A similar time series for a smaller homogeneous climate zone (in this case the South Coast Drainage Area of southern California, where the authors reside) gives us a sense of the duration of drought. This region has been

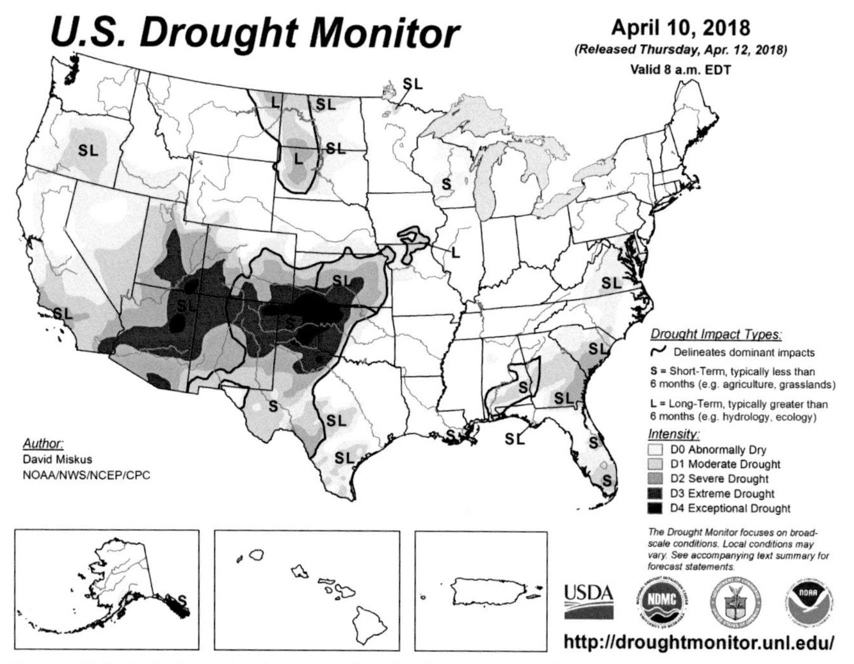

Figure 2.1 U.S. Drought Monitor for April 10, 2018. *http://droughtmonitor.unl.edu/ (accessed 12.04.18).*

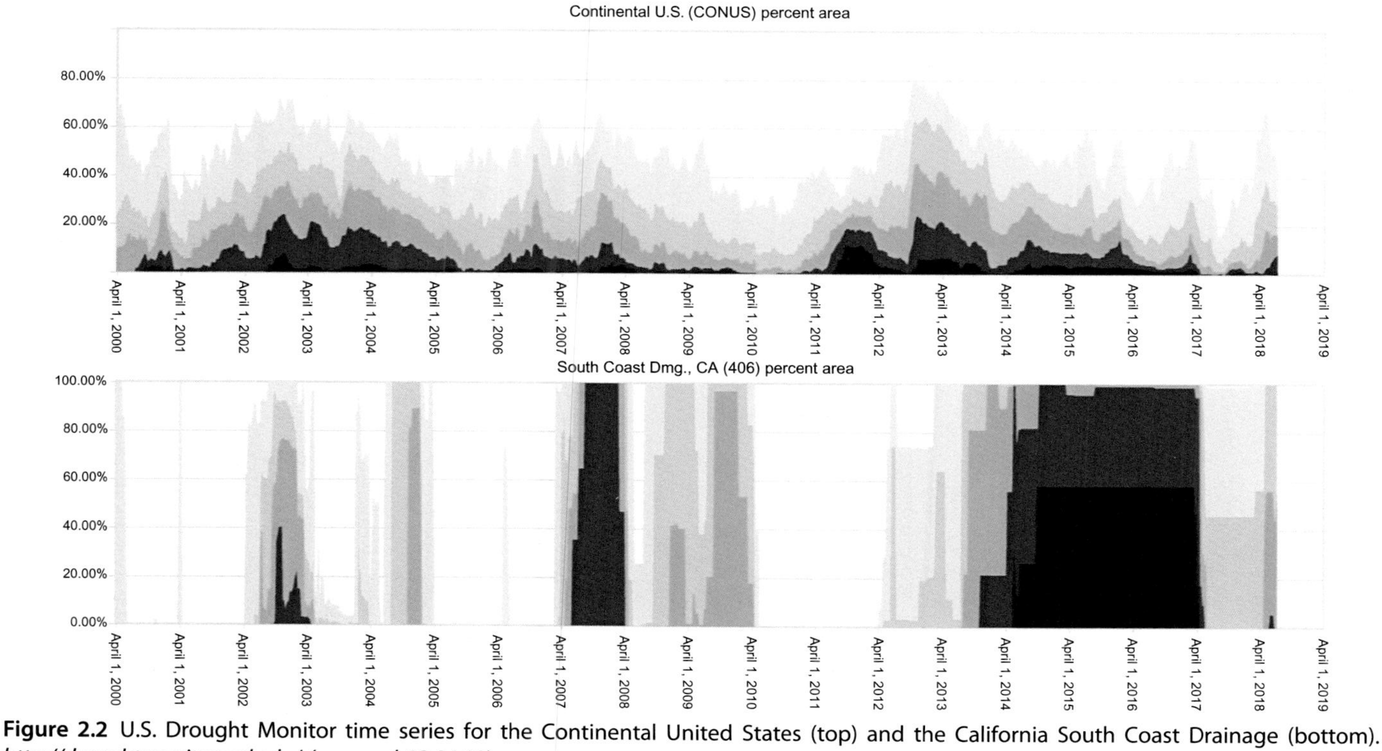

Figure 2.2 U.S. Drought Monitor time series for the Continental United States (top) and the California South Coast Drainage (bottom). *http://droughtmonitor.unl.edu/ (accessed 12.04.18).*

in some state of continuous drought since 2012, with a remarkable run of exceptional drought (D4) stretching from 2014 through early 2017, when an atmospheric river event brought some relief to the area. Atmospheric rivers are narrow bands of very moist air that form over the oceans. When they move over land, they can cause intense precipitation.

In addition to duration, severity, and extent, timing is an important but nebulous component of drought early warning. The timing aspect is highly context-sensitive. For example, depending on the growing stage of the crop, late-season dry conditions could spell disaster (if the plants are still in the grain-filling stage), or a benefit to farmers—if this growth is complete, since late-season dry conditions make it easier to harvest and store grain without waste and spoilage. In an area with a well-developed water management and storage system, an intense storm might simply provide a beneficial replenishment of reservoirs and aquifers. The same storm in a country without such infrastructure might result in widespread flooding, property loss, and even loss of life.

2.5 Impact-based sectoral or disciplinary definitions of drought

Definitions of droughts also vary according to the perspectives offered by different disciplines (Wilhite, 1993). Different disciplines can offer unique and important ways of examining, quantifying, and evaluating drought impacts. Meteorological, agricultural, hydrologic, and socioeconomic frameworks are common bases for drought evaluations. Typically, these impacts affect three principal sectors: economic, environmental, and social. Table 2.3, based on Wilhite's (1993) seminal paper, lists some of the primary impacts associated with these sectors. Economic impacts include direct agricultural losses, losses from related industries such as forestry and fisheries, and losses from the recreation, transportation, banking, and energy industries. A notable set of recent economic losses occurred in the United States in 2012, when total economic impacts exceeded 30 billion dollars (Guha-Sapir et al., 2017).

Less direct, but potentially very damaging, impacts can include disruptions in food supply and increases in food prices. Such impacts can, in fact, be felt far afield. For example, United States, Asian, and Australian droughts between 2008 and 2012 helped create a jump in global food prices, which, in turn, helped increase prices in East Africa (Davenport and Funk, 2015). Famine is ultimately caused by an inability to purchase

Table 2.3 Classification of drought-related impacts.

Problems sectors	Impacts
Economic	Loss from crop production
	annual and perennial crop losses; damage to crop quality
	reduced productivity of cropland (wind erosion, etc.)
	insect infestation
	plant disease
	wildlife damage to crops
	Loss from dairy and livestock production
	reduced productivity of rangeland
	forced reduction of foundation stock
	closure/limitation of public lands to grazing
	high cost/unavailability of water for livestock
	high cost/unavailability of feed for livestock
	high livestock mortality rates
	increased predation
	range fires
	Loss from timber production
	forest fires
	tree disease
	insect infestation
	impaired productivity of forest land
	Loss from fishery production
	damage to fish habitat
	loss of young dish due to decreased flows
	Loss of national economic growth, retardation of economic development
	income loss for farmers and others directly affect
	Loss from recreational businesses
	Loss to manufacturers and sellers of recreational equipment
	Increased energy demand and reduced supply because of drought-related power curtailments
	Costs to energy industry and consumers associated with substituting more expensive fuels (oil) for hydroelectric power
	Loss of industries directly dependent on agricultural production (e.g., machinery and fertilizer manufacturers, and food processors)
	Decline in food production/disrupted food supply
	increase in food prices
	increased importation of food (higher costs)

(Continued)

Table 2.3 (Continued)

Problems sectors	Impacts
	Unemployment from drought-related production declines
	Strain on financial institutions (foreclosures, greater credit risks, capital shortfalls, etc.)
	Revenue losses to federal, state, and local governments (from reduced tax base)
	Revenues to water supply firms
	revenue shortfalls
	windfall profits
	Loss from impaired navigability of streams, rivers, and canals
	Cost of water transport or transfer
	Cost of new or supplemental water resource development
Environmental	Damage to animal species
	wildlife habitat
	lack of feed and drinking water
	disease
	increased vulnerability to predation (e.g., from species concentration near water)
	Wind and water erosion of soils
	Damage to fish species
	Damage to plant species
	Water quality effects (e.g., salt concentration)
	Air quality effects (dust, pollutants) visual and landscape quality (dust, vegetative cover, etc.)
Social	Food shortages (decreased nutritional level, malnutrition famine)
	Loss of human life (e.g., food shortages and heat)
	Public safety from forest and range fires
	Conflicts between water users
	Health-related low flow problems (e.g., diminished sewage flows and increased pollutant concentrations)
	Inequity in the distribution of drought impacts/relief
	Decreased living conditions in rural areas
	Increased poverty
	Reduced quality of life
	Social unrest and civil strife
	Population migration (rural to urban areas)

Source: Recreated by Shukla and Funk (2019) based on Wilhite, D.A., 1993. The enigma of drought. In: Drought Assessment, Management, and Planning: Theory and Case Studies. Springer, pp. 3–15 original; modified from Wilhite, D.A., 1993. The enigma of drought. In: Drought Assessment, Management, and Planning: Theory and Case Studies. Springer, pp. 3–15.

food (Sen, 1981). High global prices, local drought, and civil war combined in 2011 to produce widespread famine in Somalia (Checchi and Robinson, 2013; Hillbruner and Moloney, 2012).

Droughts can also have profound impacts on ecosystems, and the cost of these impacts, like ecosystem services, can be hard to quantify. Drought can impact animal species by affecting wildlife habitats, food supplies, drinking water, disease, and predation. Droughts and elevated temperatures, especially when they occur repeatedly, can have extremely damaging impacts on plants. For example, repeated droughts and dry conditions have led to widespread tree mortality in California, with the U.S. Forest Service estimating 129 million dead trees from 2010 to 2017 (Fig. 2.3).

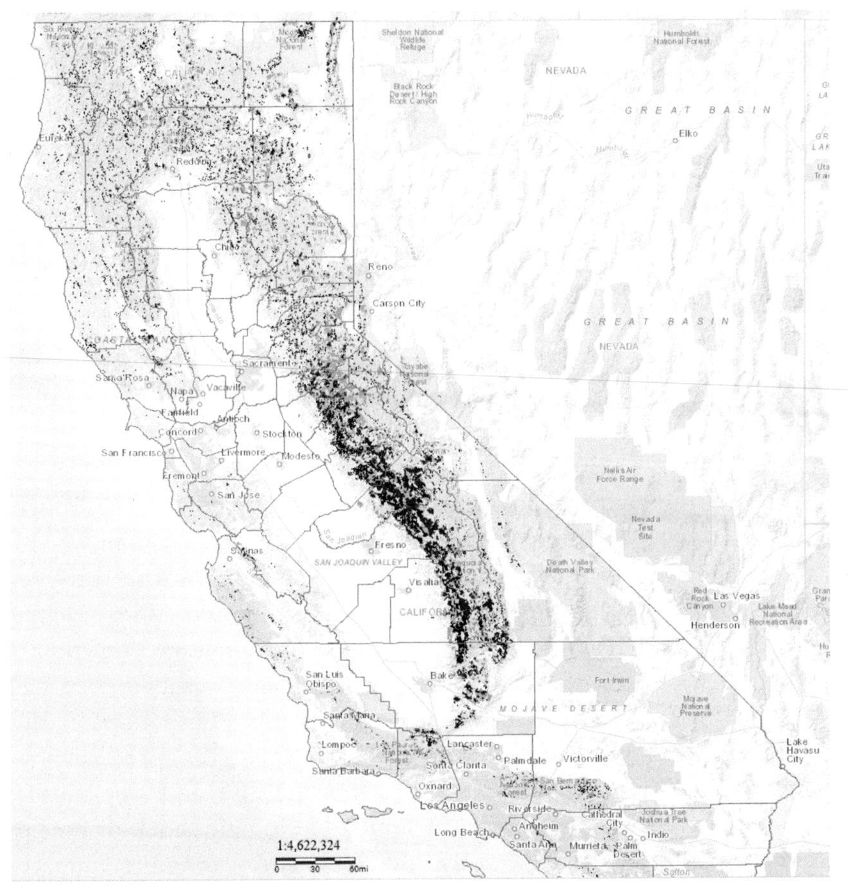

Figure 2.3 2012–17 California tree mortality. *http://egis.fire.ca.gov/TreeMortalityViewer/ (accessed 09.04.18.).*

Droughts, finally, can have profound societal impacts. At the individual or household level, droughts can lead to either chronic malnutrition, acute malnutrition, or sometimes even famine. In some cases, extremely warm temperatures alone can lead to death. For example, in 2015, a heat wave led to more than 2500 deaths, as India and Pakistan experienced deadly heat waves that exceeded 45°C (113°F) (Wehner et al., 2016). In very warm places without air conditioning, relatively modest increases in air temperatures can push the level of warmth to dangerous levels. In India, increases in temperature-related mortality seem likely associated with climate change (Mazdiyasni et al., 2017). In later chapters, we discuss how these Indian climate extremes are probably related to El Niño and may be at least partially predictable.

Many different disciplines all have valid and useful definitions of drought, leading to meteorological, agricultural, hydrological, and socioeconomic definitions of droughts. Droughts may be defined using meteorological definitions, such as when a location receives less than 200 mm (~8 in.) of rainfall during a 3-month period. Agricultural thresholds may also be used, such as when weather conditions are likely to produce a 20% reduction in yields. Agricultural outcomes are often very sensitive to the timing and location of water and heat stress. Crops are most sensitive when they switch their biological growth priorities to focus on "grain filling" because they stop adding green biomass and focus on growing reproductive grains—like corn and soybeans. Spatially, the location of droughts is critical to agriculture. Only 12% of the Earth's ice-free land surface is devoted to agriculture (Ramankutty et al., 2008).

Another important component to analyze is hydrology. A drought might arise when a river's rate of streamflow falls below a 1-in-10-year mark, while a reservoir falls beneath a certain minimum storage, or when a city exhausts its water supplies, like Cape Town, South Africa reaching their "Day Zero." Hydrologic droughts can be a complicated function of groundwater reserves, water management, and storage capabilities, as well as snowpack levels and the detailed specifics of the distribution of daily precipitation events.

The socioeconomic context in which we all live adds yet another potential disciplinary perspective. A similar hydrologic shock or precipitation deficit may have vastly different socioeconomic impacts, depending on the context, as illustrated in the next section of this chapter.

2.6 Contrasting recent U.S. and southern African droughts

To help clarify some of the concepts described here, we briefly consider and contrast the massive 2012 drought in the United States (Rippey, 2015) and the extensive 2015/16 drought across almost every country in southern Africa (Archer et al., 2017). In the central United States, the 2012 March to August (Fig. 2.4A) and June to August (Fig. 2.4B) air temperatures were some of the highest on record and precipitation totals were among some of the lowest, similar in magnitude to the dust bowl years of 1934 and 1936 (Rupp et al., 2013). Air temperatures and precipitation tend to be inversely correlated; rainfall deficits can reduce evaporation, leading to increases in surface temperatures. In 2015/16, El Niño-related climate impacts resulted in one of the worst droughts in 35 years across southern Africa (FEWS NET, 2016). In both cases, these severe droughts greatly reduced agricultural crop production.

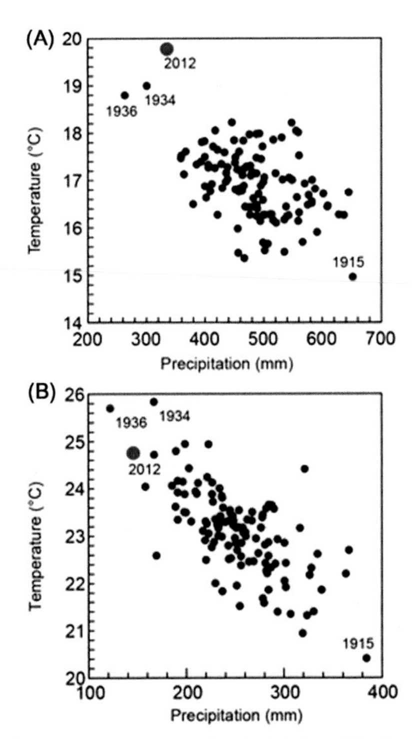

Figure 2.4 Central U.S. temperatures and rainfall for (A) the entire March–August growing season and (B) summer June–August. *Figure 1 from Rupp, D.E., Mote, P.W., Massey, N., Otto, F.E., Allen, M.R., 2013. Human influence on the probability of low precipitation in the central United States in 2012. Bull. Am. Meteorol. Soc. 94, S2.*

In the United States (Rippey, 2015), corn yields dropped dramatically. According to the U.S. Government's official disaster designations (Fig. 2.5), 5168 out of 6145 counties (84%) in the contiguous United States were considered to have experienced primary drought disasters. Estimated drought losses included ~4 billion bushels of corn, about 170 million bushels of soybeans, and 87 million bushels of sorghum, with an economic value of more than 30 billion dollars (Rippey, 2015). These crop production losses resulted in a dramatic jump in U.S. corn prices (Fig. 2.6). Unadjusted (nominal) wholesale corn prices jumped from about 180 dollars in 2010 to 300 dollars in 2012.

While the U.S. droughts in 2011 and 2012 were associated with a La Niña event, the severe southern African drought of 2015/16 was associated with one of the strongest El Niño events on record (Funk et al., 2016, 2017). The timing and pattern of the 2015/16 drought was particularly harmful to crop (maize/corn) production. Mid-season (December–February) rains failed just as most corn plants were entering their germinating or grain-filling stages. Spatially, the extent of the 2015/16 drought encompassed both the highly productive maize triangle in eastern South Africa, as well as the farms of millions of subsistence farmers in Zimbabwe, Mozambique, southern Madagascar, Malawi, and Zambia.

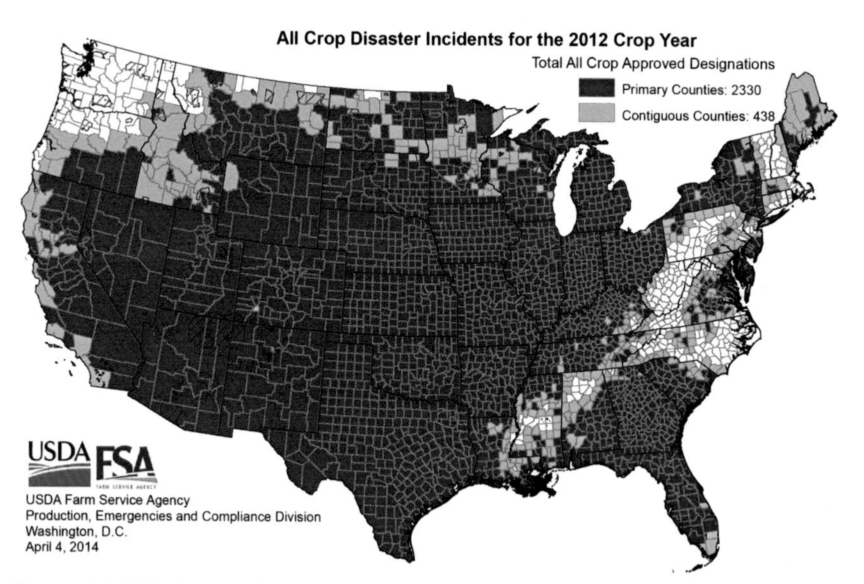

Figure 2.5 U.S. disaster designations—August 2012. *https://www.fsa.usda.gov/Assets/USDA-FSA-Public/usdafiles/Disaster-Assist/disaster_map_cropyr_2012.pdf.*

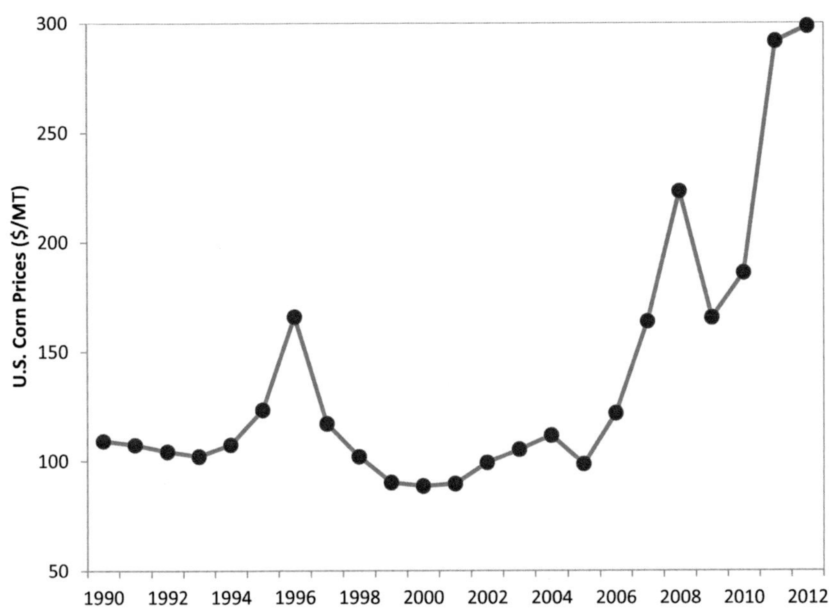

Figure 2.6 Unadjusted (nominal) U.S. corn prices. *Credit: Funk, based on data from the World Bank.*

These shocks led to a severe (~20%) regional crop-production deficit, and a corresponding large jump in cereal prices.

The 2012 U.S. and 2015/16 southern African droughts were broadly similar in magnitude and agricultural impacts—both were 1-in-20-to-30-year droughts that substantially reduced regional crop production, driving up cereal prices. The socioeconomic impacts of these droughts were vastly different. While estimates of the 2012 U.S. droughts were substantial in terms of total magnitude, ~1% of the U.S. Gross Domestic according to economic analysts[5] and about 34 billion dollars according to estimates of billion-dollar disasters from the National Center for Environmental Information,[6] the economic impact on most U.S. households was limited. Most people in the United States, however, are not involved, even indirectly, with work activities that rely on agriculture. Middle-income families in the United States spend about 13% of their income on food, while the poorest 20% of Americans spend about 35%. This tendency for poorer

[5] http://www.bloomberg.com/news/2012-11-12/u-s-drought-may-cut-gdp-by-one-percentage-point-deutsche-says.html

[6] https://www.ncdc.noaa.gov/billions/time-series

households to spend more of their income on food is known to economists as Engel's Law, so named after the 19th-century statistician Ernst Engel. In the United States the actual cost of food in a supermarket is weakly related to commodity prices, since shipping, manufacturing, and advertising typically account for most of the production costs. So, even though the 2012 drought was very large by meteorological, agricultural, and hydrological criteria, its socioeconomic impacts were limited.

The socioeconomic impacts of the southern African drought offered a stark contrast. Exceptionally warm El Niño conditions produced widespread drought and reduced streamflow and reservoir levels across much of southern Africa (Funk et al., 2016, 2017, 2018). These rainfall deficits had both primary and secondary impacts on poor households. In many cases, and immediate impact was the reduction in both household food supplies, as well as household incomes obtained from selling produce or working on nearby farms. A secondary, but equally serious problem, arose from the increase in regional food prices, thereby illustrating how droughts can trigger indirect effects that reverberate across international markets. Poor farmers in places like Malawi faced increased food insecurity, in part due to the spatially remote failure of farms located in South Africa's maize triangle—by far the largest single source of corn in southern Africa.

Fig. 2.7 shows retail maize prices—how much a poor household would pay for a kilogram in two local markets located in Lunzu and Nsanje, Malawi. At both locations, prices were more than twice the values from years past. When interpreting these numbers, it is important to keep in mind two facts. First, the poorest (lowest 20%) of households in countries such as Malawi typically spend most of their money each month on food (60% or even 70% of their household income). Second, the poorest households in poor countries have extremely limited economic purchasing power. When food prices double, these households have very limited means to make up the difference. During the 2015/16 southern African drought, economic shocks and associated food price increases pushed 16 million people into severe food insecurity (Funk et al., 2018).

2.7 Chapter review

In this chapter, we have examined various definitions of drought, as well as the challenges and opportunities provided by the particular and complex nature of droughts.

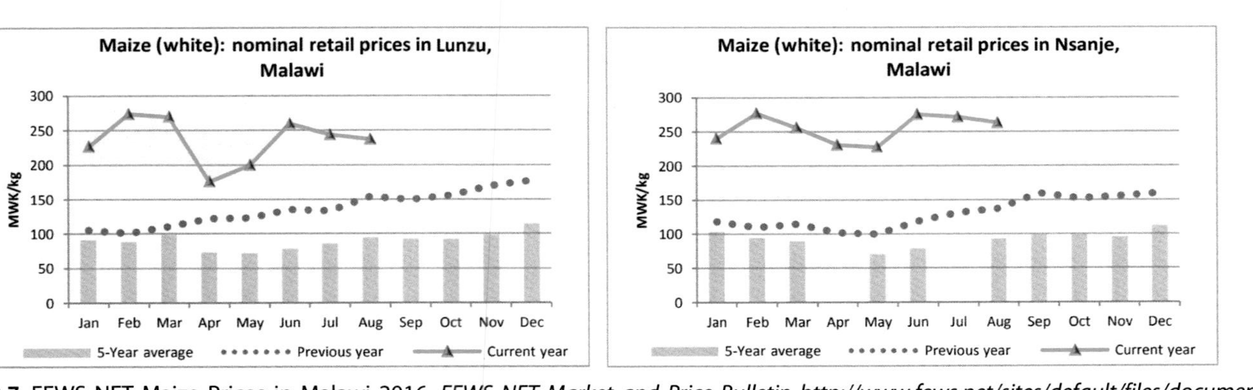

Figure 2.7 FEWS NET Maize Prices in Malawi 2016. *FEWS NET Market and Price Bulletin http://www.fews.net/sites/default/files/documents/reports/Southern_Africa_2016_09_PB.pdf (accessed 12.04.18).*

Droughts are slow-onset disasters that always involve a shortage of water. Yet while they are an extremely common natural disaster, droughts can be hard to identify and predict, in part because they can impact people, ecosystems, and economies in many different ways (Fig. 2.8). Ultimately, however, droughts always involve an interplay of water supply and water demand, and their magnitude can typically be categorized in terms of intensity, duration, and extent. Timing may play a critical role as well. For example, a brief drought during the peak of a corn (maize) growing season may limit the supply of moisture to crops just when the demand for water to grow the size of the corn kernels is at a maximum. So, a relatively short drought in a limited spatial domain might have a large impact if this drought arose in a key growing area at the peak of

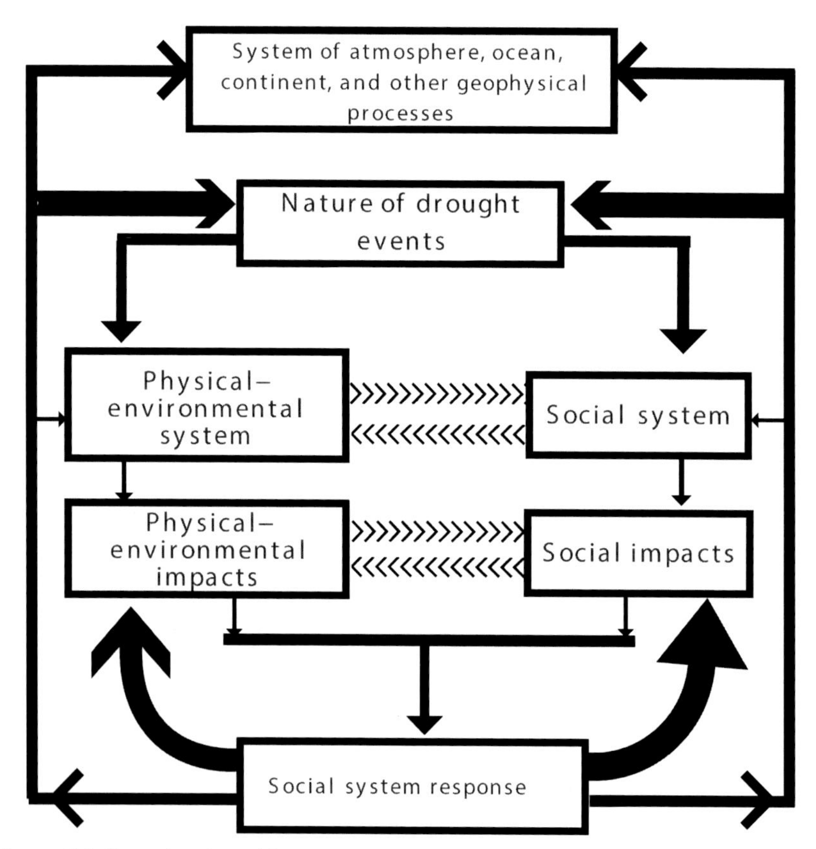

Figure 2.8 Drought viewed in a systems context. *Katie O'Brien, Recreation of Figure 2 from Wilhite, D., Glantz, M., 1985. Understanding the Drought Phenomenon: The Role of Definitions, vol. 10. pp. 111–120.*

grain filling or germination. Human drought impacts, furthermore, always arise in a human socioeconomic landscape. Contrasting recent severe droughts in the United States and southern Africa, we highlighted the very different food security outcomes associated with similar large water deficits and agricultural impacts.

In the next chapter, we explore, more completely, some current state-of-the-science DEWS.

References

Archer, E.R.M., Landman, W.A., Tadross, M.A., Malherbe, J., Weepener, H., Maluleke, P., et al., 2017. Understanding the evolution of the 2014–2016 summer rainfall seasons in Southern Africa: key lessons. Clim. Risk Manage. 16, 22–28.

Checchi, F., Robinson, W.C., 2013. Mortality Among Populations of Southern and Central Somalia Affected By Severe Food Insecurity and Famine During 2010–2012. 87 pp.

Davenport, F., Funk, C., 2015. Using time series structural characteristics to analyze grain prices in food insecure countries. Food Sec. 7, 1055–1070.

FEWS NET, 2016. Southern Africa—Illustrating the Extent and Severity of the 2015-2016 Drought. pp. 8.

Funk, C., et al., 2016. Assessing the contributions of local and east Pacific warming to the 2015 droughts in Ethiopia and Southern Africa. Bull. Am. Meteorol. Soc. 97, S75–S80.

Funk, C., et al., 2017. Anthropogenic enhancement of moderate-to-strong El Niños likely contributed to drought and poor harvests in Southern Africa during 2016. Bull. Am. Meteorol. Soc. 37.

Funk, C., et al., 2018. Examining the role of unusually warm Indo-Pacific sea surface temperatures in recent African droughts. Q. J. R. Meteorolog. Soc. 144, 360–383.

Guha-Sapir, D., Hoyois, P., Wallemacq, P., Below, R., 2017. Annual Disaster Statistical Review 2016—The Numbers and Trends. Centre for Research on the Epidemiology of Disasters (CRED).

Hillbruner, C., Moloney, G., 2012. When early warning is not enough—lessons learned from the 2011 Somalia Famine. Global Food Secur. 1, 20–28.

Mariotti, A., et al., 2013. Advancing drought understanding, monitoring, and prediction. Bull. Am. Meteorol. Soc. 94, ES186–ES188.

Mazdiyasni, O., et al., 2017. Increasing probability of mortality during Indian heat waves. Sci. Adv. 3.

Mo, K.C., Shukla, S., Lettenmaier, D.P., Chen, L.C., 2012. Do Climate Forecast System (CFSv2) forecasts improve seasonal soil moisture prediction? Geophys. Res. Lett. 39.

Pulwarty, R.S., Sivakumar, M.V., 2014. Information systems in a changing climate: early warnings and drought risk management. Weather Clim. Extremes 3, 14–21.

Ramankutty, N., Evan, A.T., Monfreda, C., Foley, J.A., 2008. Farming the planet: 1. Geographic distribution of global agricultural lands in the year 2000. Global Biogeochem. Cycles 22, n/a-n/a.

Rippey, B.R., 2015. The U.S. drought of 2012. Weather Clim. Extremes 10, 57–64.

Rupp, D.E., Mote, P.W., Massey, N., Otto, F.E., Allen, M.R., 2013. Human influence on the probability of low precipitation in the central United States in 2012. Bull. Am. Meteorol. Soc. 94, S2.

Sen, A., 1981. Poverty and Famines: An Essay on Entitlement and Deprivation. Oxford University Press.

Sheffield, J., Wood, E.F., 2012. Drought: Past Problems and Future Scenarios. Routledge.

Shukla, S., Sheffield, J., Wood, E., Lettenmaier, D., 2013. On the sources of global land surface hydrologic predictability. Hydrol. Earth Syst. Sci. Discuss. 10, 1987–2013.

Svoboda, M., et al., 2002. The drought monitor. Bull. Am. Meteorol. Soc. 83, 1181–1190.

Wehner, M., Stone, D., Krishnan, H., AchutaRao, K., Castillo, F., 2016. The deadly combination of heat and humidity in India and Pakistan in Summer 2015. Bull. Am. Meteorol. Soc. 97, S81–S86.

Wilhite, D.A., 1993. The enigma of drought. Drought Assessment, Management, and Planning: Theory and Case Studies. Springer, pp. 3–15.

Wilhite, D., Glantz, M., 1985. Understanding the Drought Phenomenon: The Role of Definitions, vol. 10. Drought Mitigation Center Faculty Publications, pp. 111–120.

Wilhite, D., Pulwarty, R.S., 2017. Drought and Water Crises: Integrating Science, Management, and Policy. CRC Press.

CHAPTER 3

Drought early warning systems

As illustrated in Table 2.3 in Chapter 2, Drought Early Warning—Definitions, Challenges, and Opportunities, drought early warning seeks to anticipate a myriad of different types of drought-related impacts. To address any one of these impact categories would almost certainly involve more expertise than any one person is likely to possess. Modern drought early warning systems (DEWS), therefore, consistent of multidisciplinary teams that examine, in detail, different types of drought risk and impact (Hao et al., 2017). As discussed in Chapter 2, Drought Early Warning—Definitions, Challenges, and Opportunities, integrated approaches to drought risk reduction stand on three pillars (Wilhite and Pulwarty, 2017): monitoring and early warning and information delivery systems, vulnerability and impact assessment, and mitigation and response. In later chapters, we will focus on the different "tools of the trade" that inform the monitoring and drought early warning pillar. Before going into these important details, however, we will describe two current state-of-the-science DEWS: the U.S. National Integrated Drought Information System (NIDIS, www.drought.gov) and the U.S. Agency for International Development (USAID)–funded Famine Early Warning Systems Network (FEWS NET). These systems were selected because they provide an excellent example of a multisectoral DEWS (NIDIS) in a data-dense region (the continental United States, CONUS) and a domain-specific system (FEWS NET), focused on food insecurity in data-sparse regions of the developing world.

3.1 The U.S. National Integrated Drought Information System

NIDIS, in partnership with the University of Nebraska-based National Drought Mitigation Center (NDMC), "jointly support or conduct impact assessment, forecast improvement, indicators and management triggers . . . and the development of portals" (Pulwarty and Sivakumar, 2014). The genesis of NIDIS began with a 2000 report, "Preparing for Drought in the 21st Century[1]." This report advocated that the United States would

[1] http://govinfo.library.unt.edu/drought/finalreport/fullreport/reportdload.htm

Drought Early Warning and Forecasting
DOI: https://doi.org/10.1016/B978-0-12-814011-6.00003-8

benefit from developing policy that promoted drought preparedness. This report also encouraged partnerships among Federal, nonfederal, and private agencies to develop appropriate tools and drought preparedness strategies. Eventually, these efforts led to Public Law 109−430, authorized by Congress in 2006. This law authorized the National Oceanic and Atmospheric Administration's (NOAA) NIDIS program. NIDIS has an interagency mandate to coordinate and integrate drought research, building upon existing federal, tribal, state, and local partnerships in support of creating a national drought early warning information system.

NIDIS "utilizes new and existing partner networks to optimize the expertise of a wide range of federal, tribal, state, local and academic partners in order to make climate and drought science readily available, easily understandable and usable for decision makers; and to improve the capacity of stakeholders to better monitor, forecast, plan for and cope with the impacts of drought[2]."

NIDIS—along with its Federal, Tribal, state, local, and private sector partners—develops leadership and partnerships, collects and integrates information, fosters and supports a research environment, and provides accurate, timely, and integrated information. While a detailed description of NIDIS is beyond the scope of this book, it is worth noting NIDIS' very well-developed multisectorial, multiregional monitoring and drought early warning capacity. At a national scale, such a system is difficult because (see footnote 2) ... "Drought in Maine looks different from drought in New Mexico. When seeking indicators of drought, a place which depends on snowpack for its annual water supply must monitor different factors from a place where liquid precipitation determines the hydrology. And local economies, resources and values influence the responses of government, business, and the public to drought prediction, conditions and aftermath".

Fig. 3.1 displays a 2019 snapshot of the main NIDIS portal.[3] Based on our definitions of drought, we can recognize descriptions of location and magnitude, expressed as a percent of area (29.7% of CONUS) and population (60.4 million people in CONUS). The text across the bottom of the page describes recent meteorological conditions, we will briefly discuss the five drought information products displayed as maps across the center of the page: the U.S. Drought Monitor, the U.S. Seasonal Drought

[2] https://www.drought.gov/drought/what-nidis

[3] www.drought.gov

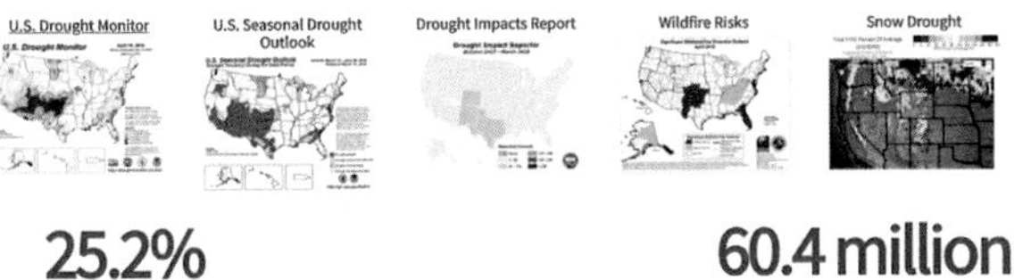

Figure 3.1 Snapshot of www.drought.gov from April 16th. *Courtesy: NIDIS www.drought.gov.*

Outlook, the Drought Impacts Report, the Wildfire Risks assessment, and the Snow Drought assessment.

As discussed in Chapter 2, Drought Early Warning—Definitions, Challenges, and Opportunities, the U.S. Drought Monitor[4] (Svoboda et al., 2002) classifies droughts into five categories: abnormally dry (D0), moderate drought (D1), severe drought (D2), extreme drought (D3), and exceptional drought (D4). These classifications are based on multiple indicators such as the Palmer Drought Severity Index (PDSI, Chapter 8: Theory—Indices for Measuring Drought Severity), Climate Prediction Center Soil Moisture Percentiles (Chapter 5: Tools of the Trade 2—Land Surface Models), U.S. Geological Survey (USGS) streamflow percentiles, Standardized Precipitation Index (SPI, Chapter 8: Theory—Indices for Measuring Drought Severity) values, and other objective drought indicator values. These quantitative metrics are augmented by local reports from more than 350 expert observers in a consultative process. The PDSI is an index, developed by W.C. Palmer in 1965, that uses rainfall and temperature data to estimate soil moisture stress. Fig. 3.2 shows a map of mid-April 2018 PDSI and a 1920–2018 time series of PDSI for Arizona. The advantage of PDSI is that it can be computed directly from climate observations, allowing calculation over a long period of record. However, the index does not take snowpack into account, an important source of water in most of the western United States. The PDSI can be slow to detect rapidly developing droughts. Fig. 3.2 (left panel) shows a map of April

[4] http://droughtmonitor.unl.edu/AboutUSDM/DroughtClassification.aspx

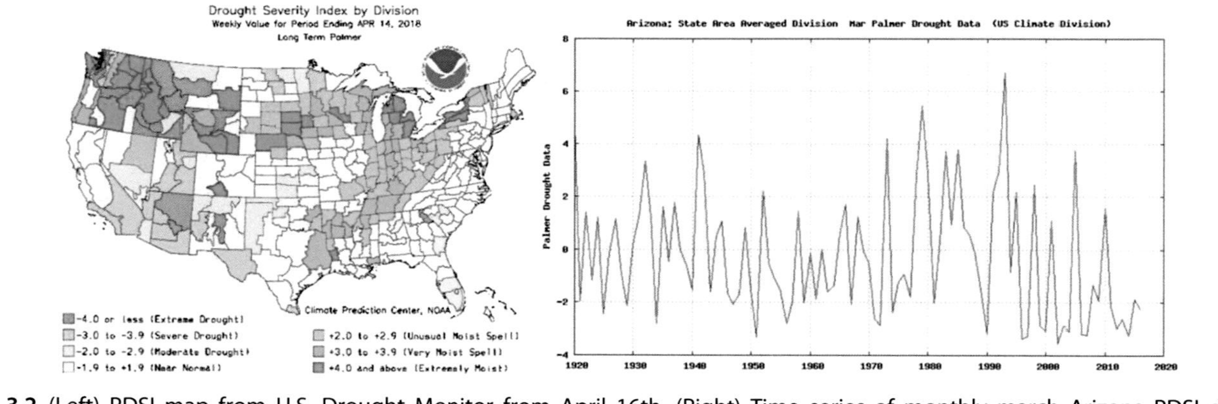

Figure 3.2 (Left) PDSI map from U.S. Drought Monitor from April 16th. (Right) Time series of monthly march Arizona PDSI data from NOAA ESRL. *ESRL*, Earth Systems Research Laboratory; *PDSI*, Palmer Drought Severity Index. *Courtesy: http://droughtmonitor.unl.edu; and https://www.esrl.noaa.gov.*

2018 PDSI for CONUS. Values of less than −2 indicate a moderate or worse drought. What we see is a tendency for dry conditions in the southern southwestern United States and relatively wet conditions over the central eastern United States. Arizona shows up as the state with the most severe dryness, and a time series of March PDSI values for Arizona (Fig. 3.2 right) indicates a series of dry spring conditions from 2011 onward. According to the PDSI, Arizona is experiencing a substantial and protracted drought.

We can confirm the PDSI-indicated dryness using two completely independent sources of information—mountain snowpack data and river streamflow observations. In operational drought early warning applications, it is generally a standard practice to examine multiple sources of information for potential convergence of evidence. All sources of information contain errors and uncertainties. When drought analysts see convergence among multiple sources of independent data sets, they can be more confident in their results. NIDIS and the U.S. Drought Monitor provide a very robust collection of independent data sets.

Fig. 3.3 left shows mountain snowpack data collected by the National Water and Climate Center, the Natural Resources Conservation Service, and the U.S. Department of Agriculture. These values are expressed as percentages of the 1981–2010 median. Across most of the central and

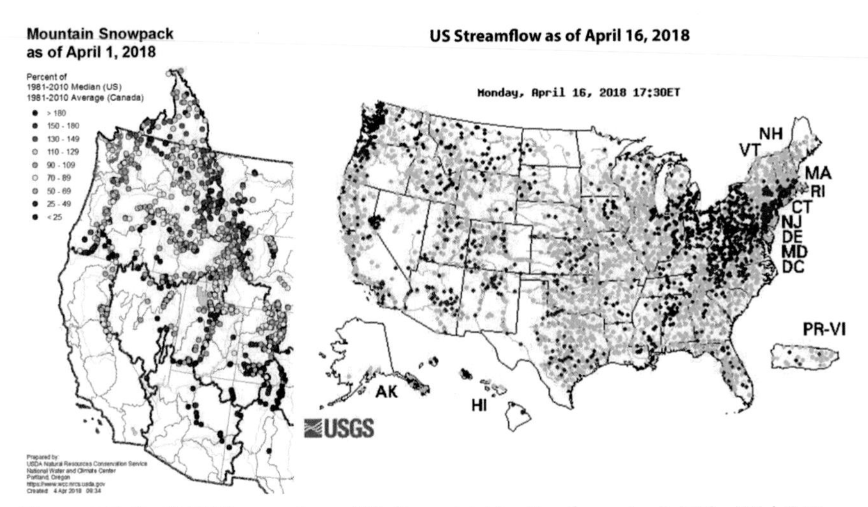

Figure 3.3 (Left) PDSI map from U.S. Drought Monitor from April 16th. (Right) Time series of monthly Arizona PDSI data from NOAA ESRL. *ESRL*, Earth Systems Research Laboratory; *PDSI*, Palmer Drought Severity Index. *Courtesy: http://droughtmonitor.unl.edu; and https://www.esrl.noaa.gov.*

southern western United States, we see below-normal snowpack levels—ranging from 70% to less than 20% of normal. In Arizona, New Mexico, southern Colorado, and southern Utah, we find very low snowpack levels. We would have to do more detailed analysis to determine whether these low snowpack levels were caused by low precipitation totals, or warm temperatures, or both, but these data, like the PDSI, suggest dry conditions in this region. In many cases, accurate snowpack observations in spring provide excellent "forecasts" of future summer water supply. In many drier, mountainous areas, lower elevation ecosystems and agriculture depend on spring and summertime snowmelt from alpine glaciers. Snowpack observations, like those shown in Fig. 3.3 left, are augmented by additional NIDIS monitoring tools,[5] such as estimates of Snow Water Equivalency (SWE). SWE estimates are based on snowpack estimates derived from a land surface model (Chapter 5: Tools of the Trade 2—Land Surface Models). These gridded estimates are typically less certain than snowpack observations but have much better spatial coverage. Advanced web-based mapping tools, like the Climate Engine,[6] can be used to provide detailed, up-to-date SWE maps.[7]

USGS streamflow observations (Fig. 3.3 right) provide yet another independent source of hydrological information. Such observations can be particularly valuable because stream gauge observations reflect not just the local precipitation, but the overall hydrologic balance in the entire watershed above our gauging location. Depending on our target application, stream gauge data may be very relevant, or not. If we are a farmer depending on rainfed agriculture, the streamflow of a river flowing 10 mi from our farm may be quite irrelevant. To a reservoir operator or hydropower company, however, such information could be extremely useful.

A fourth source of information provided by NIDIS, and particularly germane to the goals of this book, is the U.S. Seasonal Drought Outlook,[8] published by the Climate Prediction Center on the third Thursday of each month. These outlooks "depict large-scale trends based on subjectively derived probabilities guided by short- and long-term statistical and dynamical forecasts." These guidelines are based on initial conditions derived from soil moisture, current snowpack, reservoir levels, and weather and climate

[5] https://www.drought.gov/drought/data-maps-tools/snow-drought
[6] https://app.climateengine.org/
[7] http://goo.gl/5fDxnm
[8] http://www.cpc.ncep.noaa.gov/products/expert_assessment/sdo_summary.php

forecasts. Climatology also plays a major role. When we are entering a dry time of year, it is very likely that dry conditions are likely to persist. Simple as this may sound, such statements are very powerful, and a major objective of effective drought early warning. As we progress through a rainy season, approaching and then passing the time of peak precipitation and providing timely and accurate assessments of the likelihood of drought recovery is an extremely important aspect of DEWS.

The U.S. Seasonal Drought Outlook begins with the observation-based U.S. Drought Monitor, focusing on those areas of moderate or worse (D1 or higher) drought. Drought early warning specialists then predict if these regions are likely to stay the same, get better, or get worse. Areas of likely new drought development are also identified.

Systems such as NIDIS, and their many collaborating partners, leverage a vast array of observational data sets and climate model predictions (based on ocean, atmosphere, and land surface models) to characterize drought and drought impacts across the developing world. While we have focused on DEWS products for the CONUS, much of the important work carried out by NIDIS is via subnational, regional DEWS,[9] regional climate centers,[10] and state-level early warning systems.[11] Systems such as NIDIS extend far beyond forecast systems (Pulwarty and Sivakumar, 2014), combining monitoring and early warning and information delivery systems, vulnerability and impact assessments, and mitigation and response planning and policy. Focusing just on the early warning component, achieving this objective requires interoperations across many spatial scales—linking watersheds and farms with global climate features such as El Niño and La Niña, and temporal scales—linking near real-time observations with antecedent conditions and forward-looking predictions. Bringing in vulnerability and impact assessments and mitigation and response planning/policy adds a much greater level of complexity, a complexity that hopefully links decision makers at local, regional, and national scales.

Systems such as NIDIS succeed and typically become more successful, because practitioners and stakeholders practice and adapt. Effective DEWS learn from their mistakes, and the DEWS participants learn over time how to better refine and communicate their information, needs, and wants.

[9] https://www.drought.gov/drought/regions/dews
[10] https://www.drought.gov/drought/regions/rcc
[11] https://www.drought.gov/drought/regions/states

The USAID FEWS NET (www.fews.net) provides another useful example of a successful DEWS. Like NIDIS, it is well-developed, complex, and effective. Unlike NIDIS, it is focused on a very particular type of drought—drought leading to severe food insecurity in the developing world. The nature of this remit also results in a tendency to work in data-sparse regions—regions without dense weather station networks, stream gauge observations, or snowpack measurements. To understand this difference, we can plot the number of monthly precipitation gauge observations used in one FEWS NET rainfall monitoring product—the Climate Hazards Center (CHC) InfraRed Precipitation with Stations (CHIRPS) archive (Funk et al., 2015c) (Fig. 3.4). In the entire United States, we find a minimum of about 5900 stations. In Africa (excluding South Africa), we find a maximum of about 3400, but this number declines to around 500 in the late 2010s.

3.2 The Famine Early Warning Systems Network

Originally formed in response to the Sahel droughts of 1984 and 1985 (Brown, 2008), FEWS NET (www.fews.net) supports the USAID

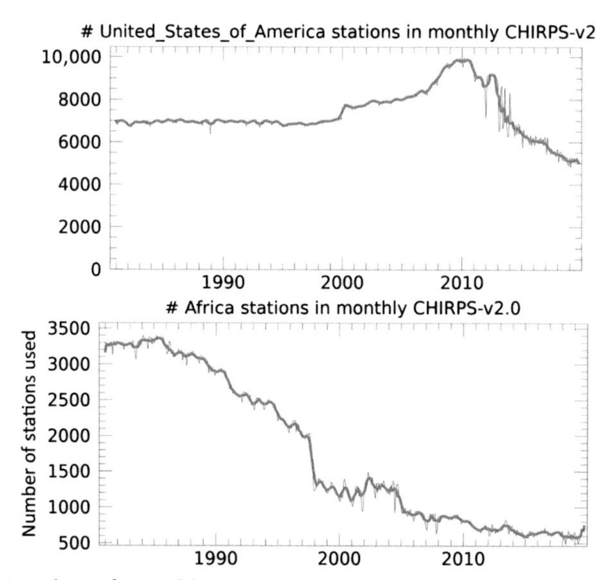

Figure 3.4 Number of monthly station observations used in the CHIRPS satellite-gauge gridded precipitation time series. (Top) United States and (bottom) all of Africa. *CHIRPS*, Climate Hazards Center InfraRed Precipitation with Stations. *Courtesy: Pete Peterson, Climate Hazards Group, University of California, Santa Barbara.*

Food for Peace (FFP) program's mission to support a world free from hunger and poverty, where all people live in dignity, peace, and security. FEWS NET is a tightly focused DEWS that also includes a highly developed analytical framework that seeks the rapid identification of emerging crisis-level (near-famine) outbreaks of acute food insecurity.[12] The timely and spatially focused alerts and outlooks provided by FEWS NET help FFP and partner agencies provide effective and early humanitarian assistance, helping to save lives and livelihoods among some of the world's most food-insecure populations. Here, we will provide an overview of FEWS NET, with a focus on the climate monitoring and prediction aspects of FEWS NET. These activities are primarily carried out by three U.S. Government science agencies: the USGS, the National Oceanic and Aeronautic Administration, and the National Aeronautics and Space Agency (NASA) (Fig. 3.5). The University of California Santa Barbara's CHC provides major science support to the USGS and also employs full-time field scientists in Africa and Central America. The U.S. Department of Agriculture, University of Maryland–led Food Security and Agriculture Consortium, and NASA/USAID SERVIR program also contribute to FEWS NET's agroclimatic monitoring efforts. FEWS NET also works closely with collaborating agencies located in Africa[13] and Europe.[14] FEWS NET's laser focus on food insecurity makes it different than most DEWS. This focus, and a more than three decade-long opportunity to refine its approach, has allowed FEWS NET to develop a very effective system for food security–related drought early warning. We will briefly describe the FEWS NET household food economy approach, current food security conditions, and the current Food Security Outlook process, as supported by the interagency FEWS NET science team. This outlook process will be the main focus of our discussion, and we will highlight areas where important scientific advances are allowing for more effective early warning.

[12] FEWS NET currently monitors 22 countries in Africa, Yemen, Afghanistan, Haiti, and Central America.

[13] A partial list of these groups would include AGRHYMET in Niamey, Niger; the IGAD Climate Predication and Applications of Centre and Regional Centre for Mapping and Resource Development in Nairobi; the SADC Climate Services Center in Gabarone, Botswana; and many national meteorological agencies.

[14] The WFP, FAO, Joint Research Centre, and European Commision.

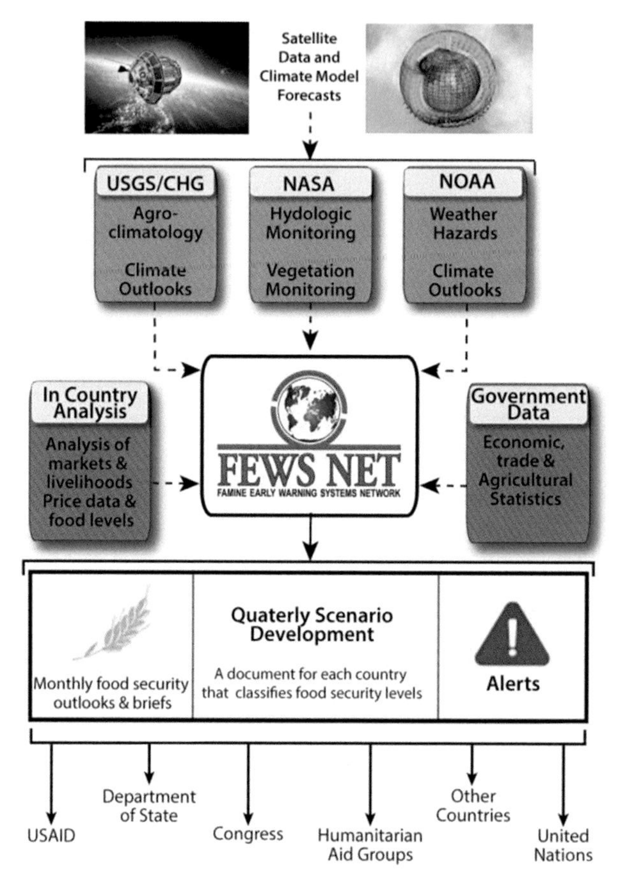

Figure 3.5 Famine Early Warning Systems diagram. *Courtesy: Chris Funk, after drawing from FEWS NET.*

3.3 Famine Early Warning Systems Network Food Security Outlooks

In general, clearly defined definitions of hazards and impacts increase the effectiveness of decision support systems. FEWS NET, in collaboration with other agencies such as the United Nations World Food Programme (WFP) and Food and Agriculture Organization (FAO), the European Commission, the African Permanent Inter-State Committee for Drought Control in the Sahel, and The InterGovernmental Authority on Development (IGAD), and nongovernmental agencies such as CARE and OXFAM, uses the international Integrated Phase Classification (IPC) system to provide consistent evidence-based assessment of severe food insecurity. The IPC is based on an analytical framework that addresses the multidimensional nature of food security issues. The IPC provides an international standard that classifies households or groups of people as facing nonstressed, stressed, crisis, emergency, or famine conditions.

In 2018 FEWS NET estimates that some 76 million people (one out of every hundred humans) in 45 countries are in IPC phase 3 or higher. Nigeria, South Sudan, Yemen, and Somalia faced the specter of famine. These totals are up 60% since 2015, and a substantial portion of these increases was associated with the severe 2015/16 El Niño and the period of severe La Niña–like climate that followed. As we will describe later, El Niños and La Niñas are associated with exceptionally warm sea surface temperatures (SSTs) in either the eastern or western tropical Pacific.

Time series of estimates of severely food-insecure East Africans (Fig. 3.6) indicate a substantial increase since 2011, and 2011 was identified as a severe drought year. At the peak of the 2011 Somali famine, some 12.6 million people are thought to have experienced prefamine conditions. In Somalia in 2011 more than 258,000 people perished, including 1 out of every 10 children under the age of 5 in the southern and central parts of the country (Checchi and Robinson, 2013; Hillbruner and Moloney, 2012). In 2017 at the peak of a similar series of La Niña–induced droughts, almost three times as many East Africans faced severe food insecurity. Somalia was particularly punished, receiving a series of poor rainy seasons in the spring of 2016, the fall of 2016, the spring of 2017, and the fall of 2017.

A core deliverable of FEWS NET is the Food Security Outlooks (FSOs). FEWS NET uses a scenario development process[15] based on livelihoods, a household food economy framework, and assumptions about future climate conditions, prices, conflict, food supplies, nutrition, labor, and other factors to develop likely subnational food security conditions for the next 1–8 months. "The strength of a scenario depends upon the development of evidence-based and well-informed assumptions about the future .... FEWS NET's analysis is livelihoods-based: all steps of scenario development are grounded in an understanding of how households in an area access food, earn income, and cope with shocks.[16]"

[15] https://www.fews.net/sites/default/files/documents/reports/Guidance_Document_Scenario_Development_2018.pdf

[16] https://www.fews.net/sites/default/files/documents/reports/Guidance_Document_Rainfall_2018.pdf

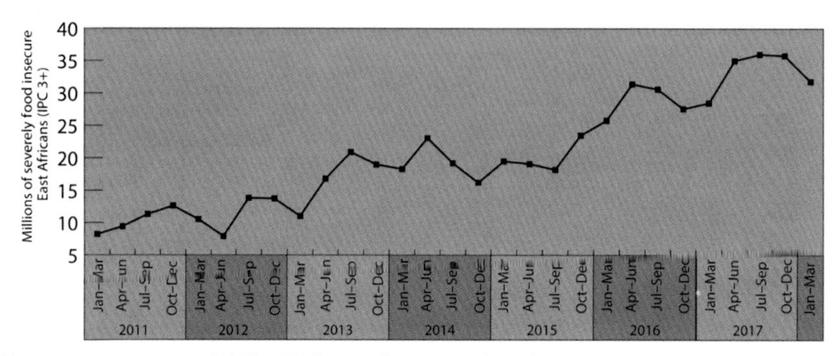

Figure 3.6 2011–18 FEWS NET/East African Food and Nutrition Working Group Food Insecure Population estimates. *FEWS NET*, Famine Early Warning Systems Network. *Courtesy: Gideon Galu, FEWS NET/Climate Hazards Group.*

FEWS NET's approach to incorporating weather and climate information typically follows a three-step progression. First, before the beginning of the rainy season, historical information is used to understand the climatology of the region, spatial patterns of exposure and vulnerability, and recent trends or changes in climate that could influence the outcome of the season. Secondly, several months before the growing season begins, various climate modes, such as the El Niño–Southern Oscillation (Hoell and Funk, 2013a; Hoell et al., 2013, 2014a,b, 2015; Korecha and Barnston, 2007), Indian Ocean Dipole (IOD), North Atlantic Oscillation (NAO) (Behera et al., 2005; Bekele-Biratu et al., 2018; Saji et al., 1999), Subtropical IOD (SIOD) (Hoell et al., 2017), or West Pacific Warming Mode (WPWM) (Funk and Hoell, 2015, 2017), may be evaluated and linked to potential climate forecasts. Third, as we approach the onset of the rainy season, climate forecasts from models, statistical analysis, and climate outlook forums are evaluated. Climate modes and forecast outlooks are examined on a monthly basis by NOAA, FEWS NET field scientists, the USGS, and CHC partners. Finally, as the season commences, FEWS NET incorporates monitoring data from remote sensing and other sources.

3.4 Multistage early warning—an Ethiopia example

Fig. 3.7 provides a schematic diagram of an effective early warning progression. We emphasize here that drought early warning may be considered as a series of increasingly strident alerts. Unlike floods, which happen rapidly, most droughts arise slowly, and typically under large-scale circulation patterns conducive to long stretches of dry weather. After more than 30 years of refinement, the FEWS NET monitoring system progresses fairly across these timescales of drought progression: from prediction (i.e., "ocean temperatures appear conducive to drought in this food-insecure region") to monitoring (looks like a drought is really happening) to detailed assessment (the harvest failed, and millions of people are likely to be in IPC crisis stage or worse).

As an example, we have used the country of Ethiopia, and we reference the El Niño–related 2015 drought event, which pushed more than 10 million people into severe food insecurity. In the left column, we list potential information sources applicable at each stage of the progression. In the central column, we list specifics pertinent to Ethiopia. On the right, we denote increasing levels of concern. As we progress through the season, our certainty and spatial specificity increase. Before the season begins, we know that Ethiopia is very food insecure, has a rapidly growing population, and has experienced an increased frequency of drought in the eastern parts of the country. In May of 2015 we see NINO3.4 SST anomalies rise above +1°C. We know that prior research has identified robust negative teleconnections associated with El Niños, and that the February–May Belg season has been poor in many places. Advancing a month, we might use land surface models, like those in the FEWS NET

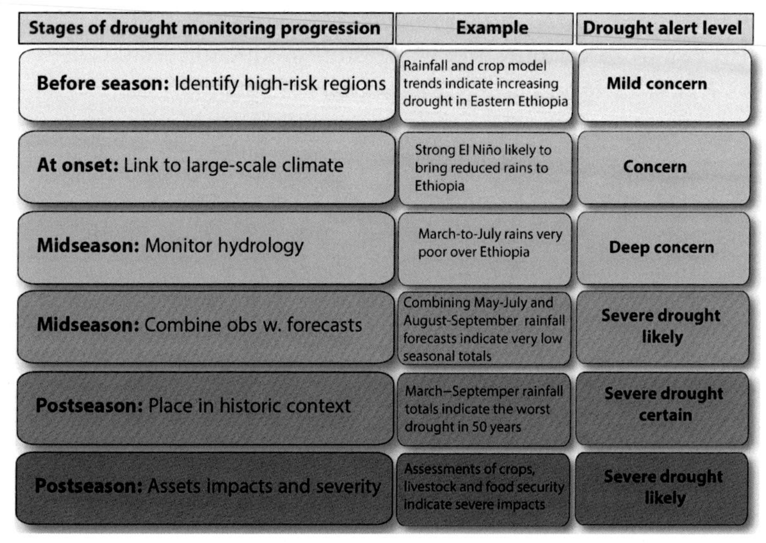

Figure 3.7 Stages of a canonical drought monitoring progression. *Courtesy: Chris Funk.*

Land Data Assimilation System (FLDAS), in conjunction with satellite precipitation fields such as the Climate Prediction Center's African Rainfall Climatology version 2 (ARC2) (Novella and Thiaw, 2013) or the CHIRPS, to examine soil moisture anomalies. Such maps would have identified some exceptionally dry conditions. Because soil moisture conditions tend to have a lot of persistence, these midseason maps are essentially predictive. Crop Water Requirement Satisfaction Index (WRSI) models provide a similar midseason snapshot of water availability. While still an active area of research, FEWS NET is working toward a capacity to combine midseason remote sensing data with forecast-based outlooks derived from weather and climate models.

It is important to note that FEWS NET depends heavily on a convergence-of-evidence approach. All sources of information are uncertain. Translating satellite-observed radiance information into accurate assessments of climatic shocks is difficult. FEWS NET, therefore, looks at multiple types and sources of remotely sensed information such as precipitation estimates, Normalized Difference Vegetation Index imagery[17] and satellite estimates of actual evapotranspiration (Senay et al., 2011, 2013, 2007). In addition to satellite information, on-the-ground reporting and station observations provide critical input to midseason assessments.

Finally, as the season draws to a close, FEWS NET early warning scientists work to refine their assessments. How *bad* might *bad* be? The Worst drought in 10, 20, 50 years? Can we use WRSI models or statistical relationships to quantify the likely crop production loss? Can we use FLDAS runoff to quantify per- capita water supplies? At the close of the growing season, high-resolution vegetation imagery provides an excellent source of spatially detailed information related to crop production and pasture conditions. Working effectively in concert, early warning products provided at each successive time period can provide increasingly accurate, specific, and actionable information.

References

Behera, S.K., Luo, J.-J., Masson, S., Delecluse, P., Gualdi, S., Navarra, A., et al., 2005. Paramount impact of the Indian Ocean dipole on the East African short rains: a CGCM study. J. Clim. 18, 4514–4530.

Bekele-Biratu, E., Thiaw, W.M., Korecha, D., 2018. Sub-seasonal variability of the Belg rains in Ethiopia. Int. J. Climatol. 38.

[17] https://earlywarning.usgs.gov/fews/product/448

Brown, M.E., 2008. Famine Early Warning Systems and Remote Sensing Data. Springer Science & Business Media.

Checchi, F., Robinson, W.C., 2013. Mortality Among Populations of Southern and Central Somalia Affected by Severe Food Insecurity and Famine During 2010–2012, 87 pp.

Funk, C., Hoell, A., 2015. The leading mode of observed and CMIP5 ENSO-residual sea surface temperatures and associated changes in Indo-Pacific climate. J. Clim. 28, 4309–4329.

Funk, C., Hoell, A., 2017. Recent climate extremes associated with the West Pacific Warming Mode. In: Wang, S.Y., Jin-Ho, Funk, C., Gillies, R. (Eds.), Climate Extremes: Patterns and Mechanisms. Wiley Press.

Funk, C., et al., 2015c. The climate hazards infrared precipitation with stations—a new environmental record for monitoring extremes. Sci. Data 2.

Hao, Z., Yuan, X., Xia, Y., Hao, F., Singh, V.P., 2017. An overview of drought monitoring and prediction systems at regional and global scales. Bull. Am. Meteorol. Soc. 98, 1879–1896.

Hillbruner, C., Moloney, G., 2012. When early warning is not enough—lessons learned from the 2011 Somalia Famine. Global Food Secur. 1, 20–28.

Hoell, A., Funk, C., 2013a. Indo-Pacific sea surface temperature influences on failed consecutive rainy seasons over eastern Africa. Clim. Dyn. 43, 1645–1660.

Hoell, A., Funk, C., Barlow, M., 2013. The regional forcing of Northern hemisphere drought during recent warm tropical west Pacific Ocean La Niña events. Clim. Dyn. 42, 3289–3311.

Hoell, A., Funk, C., Barlow, N., 2014a. The Forcing of Southwest Asia Teleconnections by low frequency Indo-Pacific sea surface temperature variability during Boreal winter. J. Clim. Available from: https://journals.ametsoc.org/doi/full/10.1175/JCLI-D-14-00344.1.

Hoell, A., Funk, C., Barlow, M., 2014b. La Niña diversity and the forcing of northwest Indian Ocean rim teleconnections. Clim. Dyn. 42, 3289–3311.

Hoell, A., Funk, C., Zinke, J., Harrison, L., 2017. Modulation of the Southern Africa precipitation response to the El Niño Southern Oscillation by the subtropical Indian Ocean Dipole. Clim. Dyn. 48, 2529–2540.

Hoell, A., Funk, C., Magadzire, T., Zinke, J., Husak, G., 2015. El Niño–Southern oscillation diversity and southern Africa teleconnections during austral summer. Clim. Dyn. 45, 1583–1599.

Korecha, D., Barnston, A.G., 2007. Predictability of June-September rainfall in Ethiopia. Monthly Weather Rev. 135, 628–650.

Novella, N.S., Thiaw, W.M., 2013. African Rainfall Climatology version 2 for famine early warning systems. J. Appl. Meteorol. Climatol. 52, 588–606.

Pulwarty, R.S., Sivakumar, M.V., 2014. Information systems in a changing climate: Early warnings and drought risk management. Weather Clim. Extremes 3, 14–21.

Saji, N.H., Goswami, B.N., Vinayachandran, P.N., Yamagata, T., 1999. A dipole mode in the tropical Indian Ocean. Nature 401, 360–363.

Senay, G.B., Budde, M., Verdin, J.P., Melesse, A.M., 2007. A coupled remote sensing and simplified surface energy balance approach to estimate actual evapotranspiration from irrigated fields. Sensors 7, 979–1000.

Senay, G.B., Budde, M.E., Verdin, J.P., 2011. Enhancing the Simplified Surface Energy Balance (SSEB) approach for estimating landscape ET: validation with the METRIC model. Agric. Water Manage. 98, 606–618.

Senay, G.B., Bohms, S., Singh, R.K., Gowda, P.H., Velpuri, N.M., Alemu, H., et al., 2013. Operational evapotranspiration mapping using remote sensing and weather

datasets: a new parameterization for the SSEB approach. JAWRA J. Am. Water Resour. Assoc. 49, 577–591.
Svoboda, M., et al., 2002. The drought monitor. Bull. Am. Meteorol. Soc. 83, 1181–1190.
Wilhite, D., Pulwarty, R.S., 2017: Drought and Water Crises: Integrating Science, Management, and Policy.

Further reading

Camberlin, P., 1997. Rainfall anomalies in the source region of the Nile and their connection with the Indian summer monsoon. J. Clim. 10, 1380–1392.
Degefu, W., 1987. Some Aspects of Meteorological Drought in Ethiopia. University Press, Cambridge.
Diro, G., Grimes, D.I.F., Black, E., 2011. Teleconnections between Ethiopian summer rainfall and sea surface temperature: part II. Seasonal forecasting. Clim. Dyn. 37, 121–131.
FEWSNET, 2015. Illustrating the Extent and Severity of the 2015 Ethiopia Drought, 7 pp.
FEWSNET, 2016. Southern Africa – Illustrating the Extent and Severity of the 2015–2016 Drought, 8 pp.
FEWSNET, 2017a. Decreases in Staple Food Prices Likely due to Humanitarian Assistance.
FEWSNET, 2017b. Large-Scale Emergencies Continue in Yemen, South Sudan, and the Horn of Africa.
FEWSNET, 2017c. Persistent Drought in Somalia Leads to Major Food Security Crisis.
FEWSNET, cited 2017d. Already Unprecedented Food Assistance Needs Grow Further; Risk of Famine Persists. Available online at: <http://www.fews.net/sites/default/files/FEWS%20NET%20Global%20FS%20Alert_20170621.pdf>.
FSNAU, 2017. FSNAU Food Security Quarterly Brief, April 2017.
Funk, C., 2016. Concerns About the Kenya/Somalia Short Rains.
Funk, C., Hoell, A., Shukla, S., Bladé, I., Liebmann, B., Roberts, J.B., et al., 2014. Predicting East African spring droughts using Pacific and Indian Ocean sea surface temperature indices. Hydrol. Earth Syst. Sci. Discuss. 11, 3111–3136.
Funk, C., Shukla, S., Hoell, A., Livneh, B., 2015a. Assessing the contributions of East African and west Pacific warming to the 2014 boreal spring East African drought. Bull. Am. Meteorol. Soc. 96, 77–81.
Funk, C., Nicholson, S.E., Landsfeld, M., Klotter, D., Peterson, P., Harrison, L., 2015b. The Centennial Trends Greater Horn of Africa precipitation dataset. Sci. Data 2.
Funk, C., Husak, G., Korecha, D., Galu, G., Shukla, S., 2016a. Below Normal Forecast for the 2017 East African Long Rains.
Funk, C., et al., 2016b. Assessing the contributions of local and east Pacific warming to the 2015 droughts in Ethiopia and Southern Africa. Bull. Am. Meteorol. Soc. 97, S75–S80.
Funk, C., et al., 2017. Anthropogenic enhancement of moderate-to-strong El Niños likely contributed to drought and poor harvests in Southern Africa during 2016. Bull. Am. Meteorol. Soc. 37.
Funk, C., et al., 2018. Examining the role of unusually warm Indo-Pacific sea surface temperatures in recent African droughts. Quart. J. Roy Meteorol. Soc. 144, 360–383.
Gill, A.E., 1980. Some simple solutions for heat-induced tropical circulation. Q. J. Roy Meteorol. Soc. 106, 447–462.

Gissila, T., Black, E., Grimes, D., Slingo, J., 2004. Seasonal forecasting of the Ethiopian summer rains. Int. J. Climatol. 24, 1345–1358.

Hoell, A., Funk, C., 2013b. The ENSO-related West Pacific Sea Surface Temperature Gradient. J. Clim. 26, 9545–9562.

Indeje, M., Semazzi, F.H., Ogallo, L.J., 2000. ENSO signals in East African rainfall seasons. Int. J. Climatol. 20, 19–46.

Jury, M., Mc Queen, C., Levey, K., 1994. SOI and QBO signals in the African region. Theor. Appl. Climatol. 50, 103–115.

Korecha, D., Sorteberg, A., 2013. Validation of operational seasonal rainfall forecast in Ethiopia. Water Resour. Res. 49, 7681–7697.

Liebmann, B., et al., 2014. Understanding recent eastern horn of Africa rainfall variability and change. J. Clim. 27, 8630–8645.

Lyon, B., DeWitt, D.G., 2012. A recent and abrupt decline in the East African long rains. Geophys. Res. Lett. 39.

Misra, V., 2003. The influence of Pacific SST variability on the precipitation over Southern Africa. J. Clim. 16, 2408–2418.

Nicholson, S.E., 1986. The nature of rainfall variability in Africa South of the Equator. J. Climatol. 6, 515–530.

Nicholson, S.E., 1997. An analysis of the ENSO signal in the tropical Atlantic and western Indian Oceans. Int. J. Climatol. 17, 345–375.

Nicholson, S.E., Kim, J., 1997. The relationship of the El Niño-Southern oscillation to African rainfall. Int. J. Climatol. 17, 117–135.

Novella, N.S., Thiaw, W.M., 2016. A seasonal rainfall performance probability tool for famine early warning systems. J. Appl. Meteorol. Climatol. 55, 2575–2586.

Ogallo, L.J., 1988. Relationships between seasonal rainfall in East Africa and the Southern Oscillation. Int. J. Climatol. 8, 31–43.

Ratnam, J.V., Behera, S.K., Masumoto, Y., Yamagata, T., 2014. Remote effects of El Niño and Modoki events on the austral summer precipitation of southern Africa. J. Clim. 27, 3802–3815.

Reason, C., Allan, R., Lindesay, J., Ansell, T., 2000. ENSO and climatic signals across the Indian Ocean basin in the global context: Part I, Interannual composite patterns. Int. J. Climatol. 20, 1285–1327.

SADC, 2016. SADC Regional Vulnerability Assessment and Analysis Synthesis Report 2016.

Segele, Z.T., Lamb, P.J., 2005. Characterization and variability of Kiremt rainy season over Ethiopia. Met. Atmos. Phys. 89, 153–180.

Segele, Z.T., Lamb, P.J., Leslie, L.M., 2009. Large-scale atmospheric circulation and global sea surface temperature associations with Horn of Africa June-September rainfall. Int. J. Climatol. 29, 1075–1100.

Shukla, S., Funk, C., Hoell, A., 2014a. Using constructed analogs to improve the skill of March-April-May precipitation forecasts in equatorial East Africa. Environ. Res. Lett. 9, 094009.

Shukla, S., McNally, A., Husak, G., Funk, C., 2014b. A seasonal agricultural drought forecast system for food-insecure regions of East Africa. Hydrol. Earth Syst. Sci. Discuss. 11, 3049–3081.

CHAPTER 4

Tools of the trade 1—weather and climate forecasts

4.1 Examples of operational drought forecasting systems

At present, there are several operational drought forecasting systems that use weather and climate forecasts to provide drought forecasts. These systems focus on different types of droughts, such as meteorological drought (based on precipitation forecasts or evaporative demand forecasts), agricultural drought (based on soil moisture or crop yield forecasts), or hydrological drought (based on runoff, streamflow, total water storage, or reservoir storage forecasts). These systems also differ in the ways that the weather and climate forecasts are used or integrated to provide drought forecasts. Nonetheless, weather and climate forecasts are a primary source for drought forecasting in all of these systems. Some of the main operational drought forecasting systems and a summary of their approaches are discussed next.

4.1.1 U.S. Climate Prediction Center's monthly and seasonal drought outlook

The U.S. National Oceanic and Atmospheric Administration's (NOAA's) Climate Prediction Center (CPC) provides monthly (Fig. 4.1A) and seasonal (Fig. 4.1B) drought outlook maps, operationally, every month.

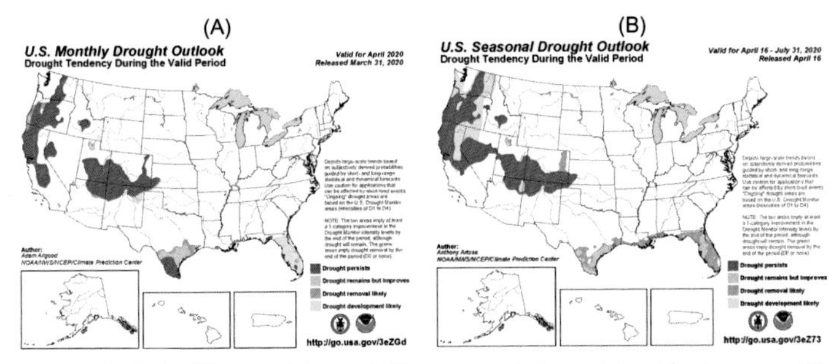

Figure 4.1 CPC's U.S. (A) monthly and (B) seasonal drought outlook maps. The maps provide an assessment of the drought conditions, mainly drought development, persistence, improvement, and recovery. The maps were accessed on April 16, 2020. *CPC*, Climate Prediction Center.

Drought Early Warning and Forecasting
DOI: https://doi.org/10.1016/B978-0-12-814011-6.00001-X

Those maps depict an assessment of the drought conditions, including drought development, progression, and recovery, based on the current drought conditions, as shown by the U.S. Drought Monitor (USDM) and weather and climate forecasts. The assessment provided is generally based on subjective (expert opinion) probability of trend in drought conditions, guided by weather and climate forecasts provided by the CPC, as well as dynamical forecasts. Like the USDM, these outlooks also benefit from the expert judgment of the drought outlook authors.

4.1.2 Famine Early Warning Systems Network food insecurity outlook

Famine Early Warning Systems Network (FEWS NET) provides assessments of food insecurity conditions (i.e., food insecurity outlooks) in some of the most vulnerable regions in the world (in about 35 + countries) to support international relief agencies in mitigating the worst impacts of food insecurity (Fig. 4.2). These outlooks are released every month. FEWS NET maps include outlooks of food insecurity conditions in the near-term (about 1–3 months in the future) and the medium-term (about 3–6 months in the future). These maps are based on several food security–related assumptions for the current time period (at the time of release of the outlooks) and for the near- and medium-term future. The future projections consider near-term and medium-term climate forecasts. The team considers forecasts of climate mode variations, such as the El Niño and Southern Oscillation (ENSO) and the Indian Ocean Dipole (IOD), since these modes influence rainfall and temperature in FEWS NET's focus regions. The outlooks also consider dynamical forecasts of rainfall, temperature, and soil moisture. For example, if a region, such as the Greater Horn of Africa, which is typically vulnerable to food insecurity, is also expected to receive below-normal rainfall, food insecurity conditions may be expected to develop or worsen there. The severity of food insecurity may depend on several other dimensions of risk, such as the livelihoods of the affected population, governance, and conflict situations. More details on the process of generating food insecurity outlooks can be found in Funk et al. (2019).

4.1.3 Miscellaneous application of weather and climate forecasts for drought forecasting

In addition to these previously mentioned drought early warning systems and several others, weather and climate forecasts are used to produce

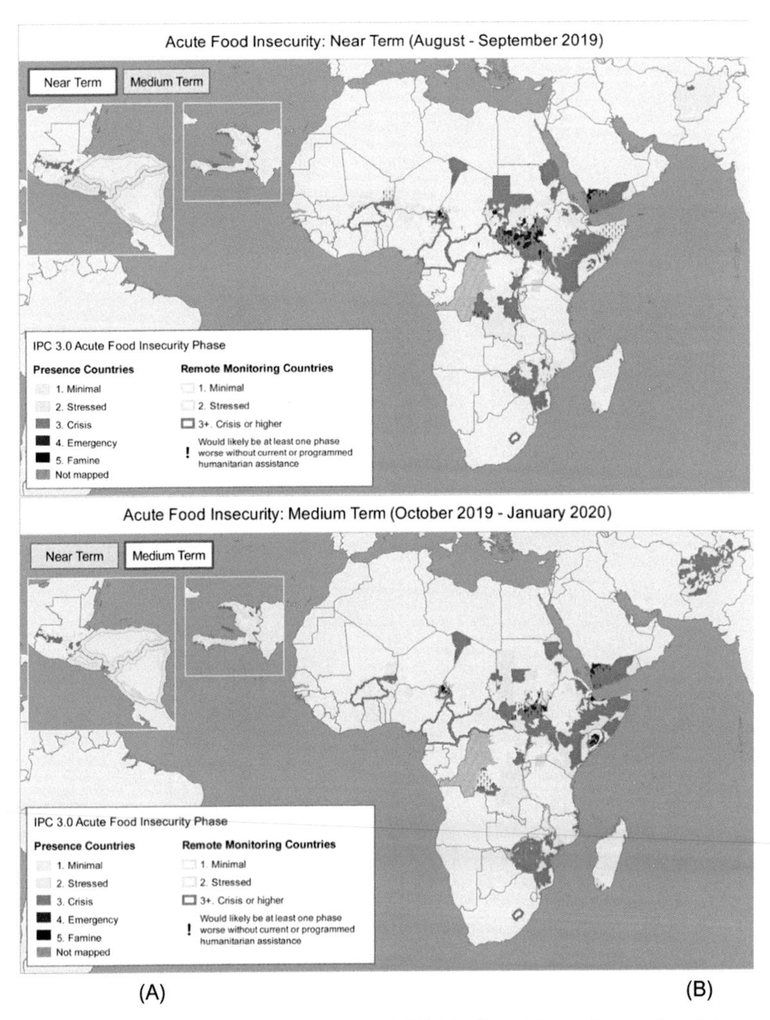

Figure 4.2 FEWS NET's (A) Near-Term and (B) Medium-Term Acute Food Insecurity outlook maps. Maps show the outlook for August–September 2019 and October 2019 to January 2020. *FEWS NET*, Famine Early Warning Systems Network.

maps of drought early indicator products. For example, CPC provides Standardized Precipitation Index forecasts based on the seasonal precipitation forecasts, and Standardized Runoff Index forecasts based on the runoff forecasts generated by forcing hydrologic models with the seasonal climate forecasts. The NASA Hydrological Forecasting and Analysis System provides forecasts of soil moisture percentile (typically considered to be an indicator of agricultural drought) for Africa and the Middle

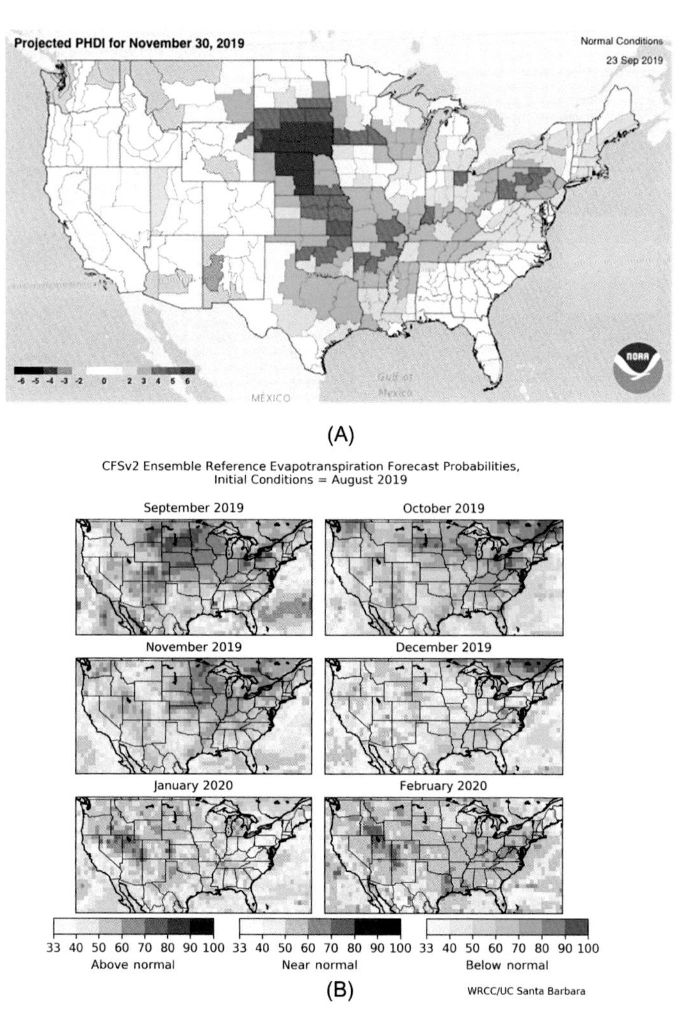

Figure 4.3 NOAA's National Centers for Environmental Information provided drought projections for the upcoming 3–12 months, based on seasonal climate forecasts. (A) PHDI forecasts for the next 2 months, assuming normal climate conditions and (B) EDDI conditions over the next 6 months, based on seasonal climate forecasts from the CFS's version 2 model. *CFS*, Climate Forecast System; *EDDI*, Evaporative Demand Drought Index; *NOAA*, National Oceanic and Atmospheric Administration; *PHDI*, Palmer Hydrologic Drought Index.

East region. The NOAA's NCEI (https://www.ncdc.noaa.gov/temp-and-precip/drought/recovery/) provides forecasts of several drought indicators based on seasonal climate forecasts (Fig. 4.3).

Due to the critical importance of weather and climate forecasts in both drought early warning and the mitigation of the adverse impacts of drought, this chapter provides a summary of fundamental details of weather and climate forecasts. The goal of this chapter is to provide a basic understanding of the process of generating weather and climate forecasts and a scientific basis for their generation. The rest of this chapter will elaborate on the (1) methods and scientific basis for generating weather and climate forecasts, (2) sources of weather and climate forecasts skill, and (3) operational weather and climate forecasting.

4.2 Methods of weather and climate forecasts

4.2.1 Climatological forecasts

The term "climatology" is often used to refer to the long-term average weather conditions in a certain place at a certain time of the year. Climatological forecasts, therefore, are simply maps of past observed climate conditions for a given region and time. Climatological forecasts are one of the simplest yet most useful (not to be confused with *skillful*) ways for providing forecasts. The assumption in using climatological forecasts is that the weather/climate during a given region and season could be similar to one or more of the weather/climate conditions during the past years at the same forecast time period and location. Climatological forecasts, although not skillful, can still be useful, as they indicate the year-to-year variability in the weather/climate conditions in a given region and time of the year. Such forecasts can be very useful, for example, in understanding when the chance of recovery for a given season becomes very high. Billions of humans live in monsoonal regions, for example, and the following narrative is very common. "When considering the drought outlook for region X, we note that the average precipitation for the next six months is very low; given this fact and the large observed rainfall deficits, we therefore conclude that the chance of extreme water stress is very high."

Typically, regions with more variable (i.e., noisy) interannual climatic conditions will have a greater need for skillful forecasts. For practical applications, climatological forecasts can also be useful in communicating the forecast uncertainty to decision-makers and can be combined with recent observations to provide skillful forecasts (depending on the time of the season) of the climate and hydrologic impacts. For example, during the middle of a rainy season, when precipitation observations since the start of the rainy season through the middle of the season are available, they can be combined with the climatological forecasts for the middle of the season through the end of the season to provide an outlook of how the precipitation over the entire season may look. In this case the skill in the observation-climatological forecasts will come from the observations through the middle of the season. If the rainy season, from the start to the middle of the season, has been poor, there will be a higher chance of an overall poor rainy season than if the season started out well. Another important practical application of climatological forecasts is that they serve as a benchmark for calculating the skill of more sophisticated weather/climate forecasts (i.e., forecasts generated using global climate models and/or statistical models informed by the initial state of the climate). Typically, the more sophisticated forecasts will beat climatological forecasts in terms of usefulness for decision-making applications. Chapter 10, Practice—Evaluating Forecast Skill, describes some techniques for evaluating forecast skill.

4.2.2 Persistence-based forecasts

Like climatological forecasts, persistence forecasts also typically serve as a benchmark for evaluating the value of sophisticated forecasts. The main difference between both types of forecasts is that persistence-based forecasts rely on observations during the recent past rather than observations from several past years. Here, the assumption is that the forecasts during the upcoming time period may be similar to what has happened in the recent past. Using the analogy of the example presented previously, the assumption would be that if a season has been rainy from the start of the season to the middle of the season, it may be rainy during the rest of the season as well. The skill of persistence forecasts varies based on the forecast lead-time. For example, persistence forecasts are typically most skillful for the next day's weather. In other words, assuming that today's weather will be the same as yesterday's is often adequate. In general, persistence forecasts are implemented as autoregressive statistical models. In a typical autoregressive model, it is assumed that the values over the next time steps depend linearly on the values in the current time step, as well as a stochastic component (inherently unpredictable component).

4.2.3 Analog forecasts

Like climatology and persistence-based forecasts, analog forecasts are also informed by past observations. However, in the case of analog forecasts, past observations are screened based on their similarities with the current conditions and/or expected future conditions. For example, during an El Niño year, it would be fair to assume that future climate conditions can be similar to the climate conditions during the past El Niño years. Here, El Niño conditions are used as a criterion to screen best matches—analogs—from the past observations. The analog forecasts can be most useful when a sufficiently large sample of past observations exists, which essentially increases the chances of finding a best match for the current or expected weather/climate conditions. Methods also exist to calculate weighted average of past analogs where weights are assigned based on the similarity of each of the analogs with the current or expected conditions.

4.2.4 Statistical forecasts

Here, we refer to statistical weather and climate forecasts that utilize a statistical method to represent the generally linear (and in some cases nonlinear) relationship between the predictand and predictors. Predictors in this case are generally current, or recent-past observations, which either statistically (e.g., estimated using correlation) or dynamically (e.g., proved by theory and/or model-based experiments) are known to have some bearing on the future weather/climate conditions. Statistical forecasts generally rely on the assumption that past relationships between the predictand and predictors will remain intact in the future. Statistical forecasts generally rely on the signal (i.e., the strength of the linear relationship) derived from the predictors and stochastic component to represent the noise in the relationship. Use of statistical forecasts is prevalent because they are numerically less expensive than numerical weather predictions and can still provide a useful level of skill for decision-making applications. They can also be explained with relative ease that makes them an attractive tool for the decision-makers. Chapter 9, Sources of Drought Early Warning Skill, Staged Prediction Systems, and an Example for Somalia, provides some examples of statistical forecast models.

4.2.5 Dynamical forecasting

Dynamical forecasts (also often known as numerical weather prediction or NWP) are the most sophisticated, and presumably, the most realistic representation of the climatic phenomenon. The history of NWP is a long one. The theoretical basis for "long-range" (i.e., several days in the future) NWP was put forth by Abbe (1901). His seminal paper posed the long-range weather forecasting as an initial value problem and proposed that future weather could be predicted using physical laws (i.e., the governing partial differential equations) starting from the initial state of climate (i.e., observed current state of climate). Dr. Edward Lorenz's work highlighted the inherent unpredictability or "chaos" in the climate system, which led to the realization of the limits of predictability, as well as the importance of improving estimates of initial conditions. Major progress has been made to improve initial conditions via satellite data sets and/or by data assimilation. Dr. Lorenz's work showed that a relatively small uncertainty (or error) existed in the estimates of initial conditions that lead to higher level of uncertainty as time passed. In other words, higher lead forecasts will have a higher level of uncertainty due to uncertainty in the initial state.

Since the early 20th century, major advances have been made in weather and climate forecasting by improvements in (1) monitoring the current state of climate, thus leading to improved knowledge of the initial state and (2) computing resources leading to higher resolution and faster calculation of the governing differential equations. Overall, these advances have led to the development of several operational weather- and climate-forecasting systems, and substantial improvements in weather and climate forecast skill. For example, in general, 3–10 days ahead weather forecast skill has been increasing by about 1 day per decade (Bauer et al., 2015).

4.3 Sources of weather and climate forecast skill

There are several sources of weather and climate forecast skill. Their relative importance in terms of the contribution to the skill changes with forecast, time, and space. The sources of skills that are most important for short-range weather forecasts are different than the most important sources for long-range and subseasonal to seasonal scale forecasts. This is mainly due to the fact that different sources of skill have different levels of persistence and periods of oscillation. For example, sea ice forcings have larger persistence than ocean surface temperature conditions, which have longer persistence than soil moisture, all of which influence future weather and climate conditions.

The first-type sources are climate and weather modes of variability, such as ENSO, Madden-Julian Oscillations, and IOD, which have a certain recurring and/or quasioscillatory frequency. Their respective phases (such as negative and positive phases) repeat after a certain quasiperiodic interval. These are commonly referred to as "modes" of variability. Their period of oscillations can range from 2 weeks to several years (typically 3–5 years). Their teleconnection to the weather and climate of regions across the globe, and at different time ranges (or lead-times), is an important source of the weather and climate forecast skill. Major attempts have been made to understand (1) how these modes of variability evolve and (2) what their overarching influences on weather and climate are. These modes are used in both a dynamical and statistical manner to provide forecasts of weather and climate. For example, ENSO, which is one of the most widely known and influential modes of variability, can affect precipitation and temperature across the globe.

The second-type information sources are the anomalies in the components of the Earth system whose persistence lasts for a similar time range as the target forecast period. For example, soil moisture anomalies can persist up to a few weeks and hence influence weather and climate during that period through the land-atmospheric feedback process. Similarly, anomalies in the ocean surface temperature can last for months and hence influence the climate over several months.

The third-type sources are the "external" radiative forcings that can be generated over different time periods and influence climate over weeks to months, or longer. Examples of these forcings include aerosols from volcanoes, carbon emissions, and cyclic and anomalous solar output. These atmospheric aerosols and greenhouse gasses modify clouds, precipitation, and the atmosphere's ability to trap and retain radiation. Typically, these sources of weather/climate forecast skill, along with slowly varying sea ice and ocean temperatures, are accounted for in forecasting systems either as "boundary values," "initial values," or a combination of both.

4.4 Summary

Weather and climate forecasts are crucially important for an effective drought early warning system. Therefore this chapter first provides examples of how they are used for drought forecasting (NOAA CPC's Seasonal Drought Outlook) and for food insecurity early warning (U.S. Agency for International Development's FEWS NET). This chapter then provides a summary of different methods to provide the forecasts, which are generally classified into two categories: statistical and dynamical methods. Finally, this chapter provides a brief description of the sources of the skill in forecasts, which form a basis for their generation and application for providing drought forecasts and usage in drought decision-making.

References

Abbe, C., 1901. The physical basis of long-range weather forecasts. Mon. Weather Rev. 29, 551–561. Available from: https://doi.org/10.1175/1520-0493(1901)29[551c:TPBOLW]2.0.CO;2.

Bauer, P., Thorpe, A., Brunet, G., 2015. The quiet revolution of numerical weather prediction. Nature 525, 47–55. Available from: https://doi.org/10.1038/nature14956.

Funk, C., et al., 2019. Recognizing the Famine Early Warning Systems Network (FEWS NET): over 30 years of drought early warning science advances and partnerships promoting global food security. Bull. Am. Meteorol. Soc. . Available from: https://doi.org/10.1175/BAMS-D-17-0233.1.

Further reading

Lorenz, E.N., 1963. Deterministic nonperiodic flow. J. Atmos. Sci. 20 (2), 130–141.

National Academies of Sciences Engineering and Medicine, 2016. Next Generation Earth System Prediction: Strategies for Subseasonal to Seasonal Forecasts.

Shukla, S., Funk, C., Hoell, A., 2014. Using constructed analogs to improve the skill of National Multi-Model Ensemble March-April-May precipitation forecasts in equatorial East Africa. Environ. Res. Lett. 9, 094009. Available from: https://doi.org/10.1088/1748-9326/9/9/094009.

CHAPTER 5

Tools of the trade 2—land surface models

5.1 Introduction

Drought monitoring benefits from estimates of current conditions (including estimates of precipitation totals, soil moisture, and streamflow) and a long-term record of these estimates to compare with current conditions. Ideally, drought monitoring would be facilitated by spatially well-distributed in situ and long-term observations of different water cycle components, but these observations often do not exist in developing countries and remote locations. In general, the records of precipitation and temperature are more prevalent than the records of water cycle components such as soil moisture and streamflow. Due to this general lack of hydrologic observations, researchers and engineers have utilized physical and/or empirical mathematical equations to predict (or estimate) streamflow from available observations of precipitation and temperature. Adequate representation of land surface processes is necessary and, in turn, the water and energy budget provided by land surface models (LSMs) helps to drive models of the global climate. This need has led to the field of land surface modeling.

The evolution of modern-day LSMs can be attributed to the need for accurately representing energy, water balance, and land-atmosphere interactions in the general circulation models (GCMs). During the advent of the GCMs, land surface would be represented by a simple leaky bucket model. This approach did not account for variability in infiltration rate due to topography and soil parameters, or the effect of vegetation resistance on evapotranspiration. Several studies identified issues with this approach, which led to the advent of more complex LSMs. Although they were initially proposed as a land surface scheme of the GCMs, the LSMs, as independent models, started to be used to run offline (i.e., when an LSM is not coupled with a GCM and is forced by prescribed atmospheric forcings) to simulate water balance components. It was this initial

Drought Early Warning and Forecasting
DOI: https://doi.org/10.1016/B978-0-12-814011-6.00005-1

independent use of the LSMs that led to their applications in hydrologic and drought monitoring and forecasting. Next, we provide a brief overview of the LSMs, a history of their usage in drought monitoring, and examples of experimental or operational drought monitors that are based on LSMs. Finally, we discuss the limitations of the LSMs and identify the issues with regard to their usage in drought monitoring.

5.2 An overview of land surface models

The recognition of the importance of the land surface in partitioning the energy balance—and as a carbon sink that is crucial for climate as a whole—led to the development of modern-day LSMs. Fig. 5.1 provides a schematic of global energy balance, which essentially accounts for the incoming and outgoing radiation and atmospheric and terrestrial fluxes. The land surface mainly plays a role in this energy balance by partitioning the net energy into sensible and latent heat fluxes. The net energy is the sum of all the upward and downward radiation entering and leaving the land surface. Partitioning of the available net energy into sensible and latent heat fluxes is dependent on the available moisture in the land

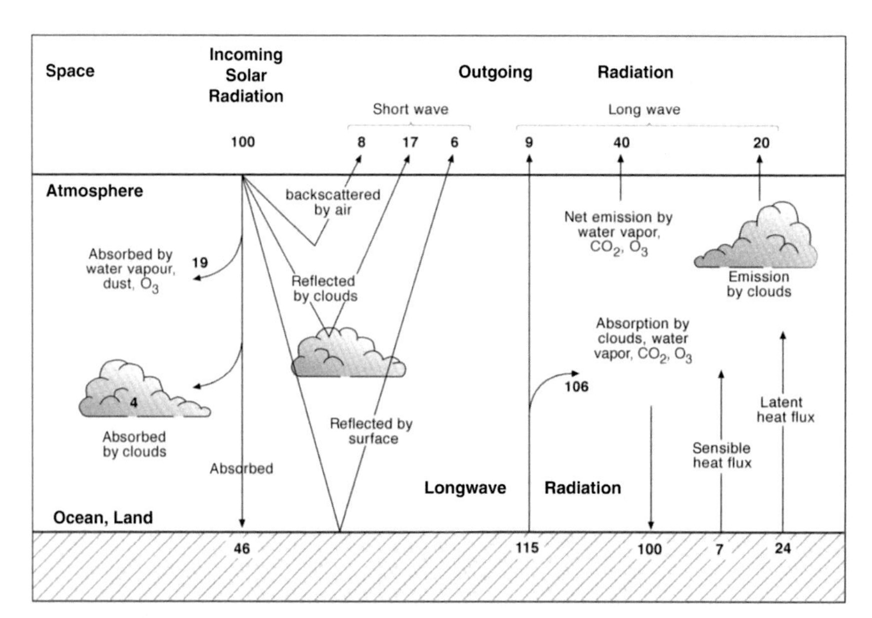

Figure 5.1 Schematic representation of global energy balance.

surface and vegetation type. An adequate estimation of the partitioning of energy is conducted by the LSMs.

At present, globally, there are several widely used macroscale LSMs that are currently used for drought monitoring purposes. This section will provide a brief overview of the typical LSM schemes. The LSMs use mathematical equations, along with physical or empirical parameters, to calculate water and energy budget at the interface of land and atmosphere. A typical LSM scheme comprises vegetation, soil, and snow, as well as a representation of the land surface's interaction with the rest of the climate to account for water and energy. Different LSM schemes often vary in how they represent those components. For example, the representation of soil profiles is often done differently, depending on the LSMs. Typically, the LSMs have three or four layers of soil with varying depths; however, there are some models with one layer of soil. The depth of the soil layers is often prescribed and/or calibrated to match certain observed hydrologic variables (e.g., streamflow). Vegetation is represented with different complexities as well. Models such as variable infiltration capacity (VIC) allow for different vegetation classes to exist within the same grid cell. The fraction of the vegetation cover in a grid cell occupied by a given vegetation class is often determined using satellite-based observations of vegetation over an extended period of time. Rooting depth is often prescribed as well. The models also often assume a static seasonal cycle of vegetation, meaning that they simulate the growth cycle of vegetation types during a given year, but that cycle does not change from one year to the next. The variables, such as leaf area index at any given time of the year, remain the same each year.

Snow representation among LSMs also varies. There are now a few LSMs that have multiple layers of snow. Typically, the input precipitation (direct rainfall or snowmelt) is first used to satisfy deficit in soil moisture, and excess water is simulated as runoff. The moisture in the soil is withdrawn through evaporation (bare land) and transpiration via the canopy (in vegetative land), or percolation to the deeper layer of substorage.

Fig. 5.2 provides a schematic of the water and energy balance, as simulated by a given grid cell of the VIC model (Liang et al., 1994). Precipitation (P) accounts for input water in a given grid cell, and E (evaporation), E_t (evapotranspiration), E_c (canopy evaporation), R (runoff), and B (baseflow) are outputs of moisture from the given grid cell. Available net energy and amount of moisture in the soil influence the

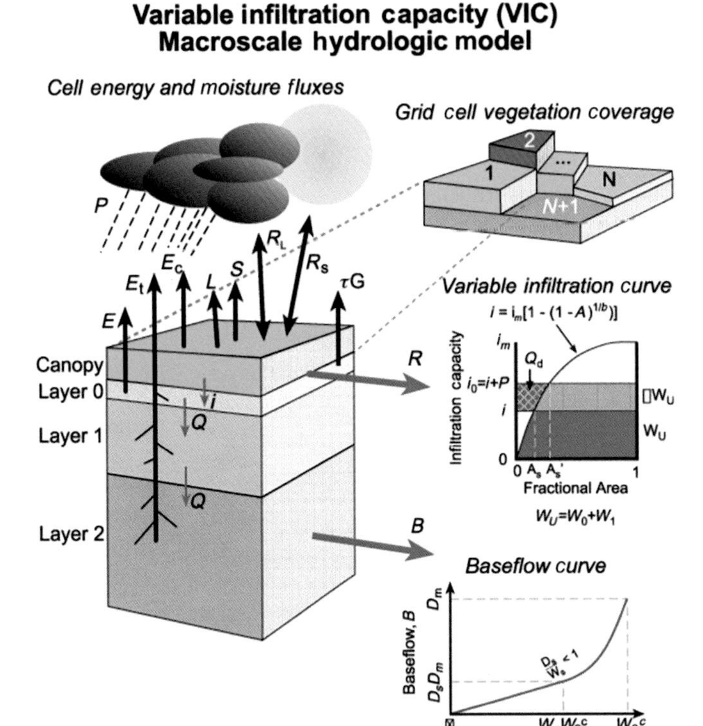

Figure 5.2 A schematic of the water and energy balance, as simulated by a given grid cell of the VIC model (Liang et al., 1994). *VIC*, Variable infiltration capacity.

amount of evaporation output, and the amount of moisture and rainfall intensity influence runoff and baseflow out of the grid cell. A VIC model does not allow for lateral movement of water from one grid cell to another; however, external models (typically known as routing models) can take the surface and subsurface runoff from the grid cells and simulate its flow through a given basin, based on the basin digital elevation model, surface roughness, etc.

5.3 Operational land surface models–based drought monitors

This section provides an overview of currently operational national- or continental-scale drought monitors that are based on LSMs.

5.3.1 National Center for Environmental Prediction's North American Land Data Assimilation System (NLDAS) drought monitor

National Oceanic and Atmospheric Administration's (NOAA's) National Center for Environmental Prediction (NCEP)'s North American Land Data Assimilation System (NLDAS) hydrologic modeling system (Mitchell et al., 2004) is one of the first operational multiple LSM-based drought monitoring systems.[1] Each day, four of the LSMs (VIC, NOAH, MOSAIC, and SAC) are run using the same atmospheric forcing data through the past day to generate simulations of soil moisture and runoff (among other water and energy balance variables) for the past day. Current soil moisture and runoff data are used for monitoring drought by comparing them with a long-term (as far back as 1979) historical distribution of those variables for each grid cell. By comparing these data with the historical distributions, current values are converted into percentiles, which indicate the severity of the current drought conditions. Total soil moisture (a sum of the soil moisture of all layers, which are different for each model) and top-one-meter soil moisture are used to provide estimates of current drought severity. Top-one-meter soil moisture is presumably used for agricultural drought monitoring, and total soil moisture is most appropriate for hydrological drought monitoring, as it takes into account the moisture in the deeper layer. This contributes to baseflow and changes relatively slower than the top layers. Fig. 5.3 depicts total soil moisture percentile, as of April 7, 2018, over the Continental United States (CONUS). The areas in drought are marked by below 30 percentile of soil moisture. The regions with the least soil moisture percentile are estimated to be in the most severe drought compared to their respective climates (in this case spanning from 1979 to present). The LSM-based drought monitors [such as NLDAS drought monitors and monitors such as Climate Prediction Center's (CPC's) Land Surface Monitoring and Prediction System[2] and Surface Water Monitor[3]] have been crucial in improving drought monitoring in the CONUS. These products are one of the data sets that are used by the U.S. Drought Monitor.[4]

[1] http://www.emc.ncep.noaa.gov/mmb/nldas/drought/

[2] http://www.cpc.ncep.noaa.gov/products/Soilmst_Monitoring/US/Soilmst/Soilmst.shtml

[3] http://www.hydro.washington.edu/forecast/monitor/index.shtml

[4] http://droughtmonitor.unl.edu

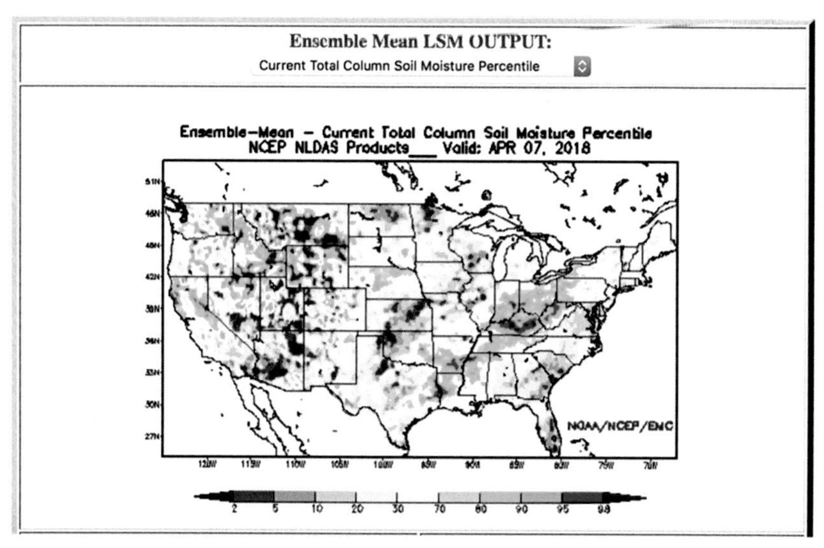

Figure 5.3 Example of NLDAS drought monitor.

5.3.2 Africa Flood and Drought monitor

Princeton University—in collaboration with UNESCO, AGHRYMET (Niger), and ICPAC (Kenya)—has produced the African Flood and Drought Monitor (AFDM).[5] AFDM is another land surface–based hydrologic monitoring system that provides estimates of flood and drought conditions in Africa (Sheffield et al., 2014). This monitoring system, too, is grid-based and uses gridded atmospheric forcing data derived from satellite, reanalysis, and in situ observations (Sheffield et al., 2006). It is based on the VIC LSM. The VIC LSM is forced with gridded atmospheric forcings and runs in real-time (with a lag of 3 days or so) to provide simulations of soil moisture, runoff, and evapotranspiration for all of Africa. These gridded, simulated values of hydrologic variables are then converted into percentiles based on the historical distributions, or historical simulations, of the respective variables. This approach of providing drought and flood severity estimates based on

[5] http://stream.princeton.edu/AWCM/WEBPAGE/interface.php

percentiles of simulated hydrologic outputs is similar to the approach used by the NLDAS. This system also provides a precipitation- and vegetation indices—based drought indicator, which can be used in conjunction with the simulated outputs for flood and drought monitoring.

One of the primary differences between the AFDM and NLDAS's drought monitor (besides the change in domain and atmospheric forcings data sets) is that the AFDM system is based on one LSM only, whereas the NLDAS drought monitor is based on multiple LSMs. Incorporating multiple LSMs provides a better estimate of uncertainties in drought monitoring (Wang et al., 2009). Different LSMs have different assumptions regarding their land surface schemes, along with their parameters. Therefore using multiple models provides a range of outputs for drought monitoring that takes model-related uncertainties into account (Fig. 5.4).

5.3.3 Global soil moisture monitoring

Thus far, there have been several attempts to provide hydrologic monitoring at a global scale. Some of those attempts have been limited to researching and exploring the potential of global-scale hydrologic monitoring using LSMs. One such attempt is described in Nijssen et al. (2014). This study developed a prototype global-scale drought monitoring technique using three LSMs: VIC, NOAH, and Sacramento. The historical simulations (used for historical distribution of hydrologic variables) were generated using Princeton's atmospheric forcing data set (Sheffield et al.,

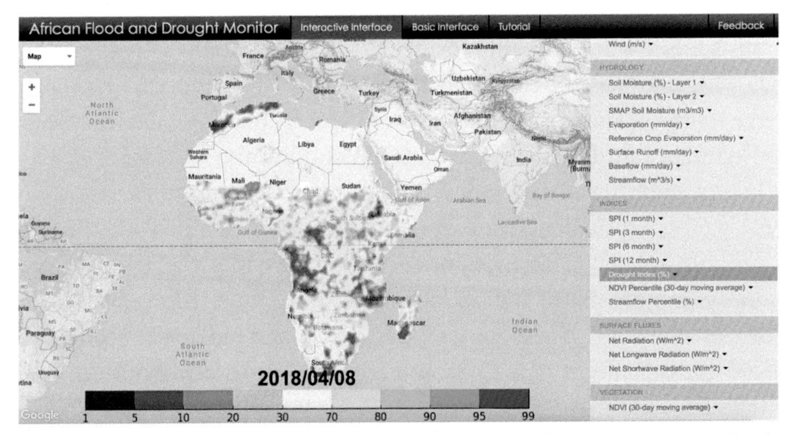

Figure 5.4 Total soil moisture percentile over Africa as of April 8, 2018, as displayed by the Princeton and partner institutes' AFDM. *AFDM*, Africa Flood and Drought Monitor.

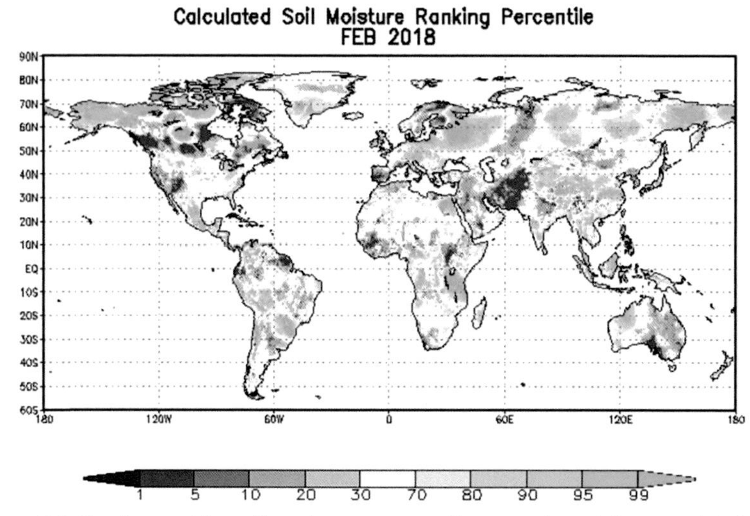

Figure 5.5 Total monthly soil moisture percentile over the globe as of February 2018, as displayed by the NOAA's CPC's Soil Moisture Monitoring system. *CPC's*, Climate Prediction Center's.

2006). Each of the LSMs was run with the same forcings. For the real-time simulations the study used a combination of bias-corrected (with respect to Princeton's atmospheric forcings) satellite precipitation National Aeronautics and Space Agency (NASA)'s TRMM (Huffman et al., 2006) and NCEP's Global Ensemble Forecasts System temperature analysis fields (Hamill et al., 2013). The study evaluated the performance of this system for monitoring major-known drought across the world and concluded that this system would be able to identify major drought events across the globe. The study also highlighted the challenge of using satellite data sets and the temporal inconsistency of forcings data set, which can lead to biased estimates of the current hydrologic conditions.

One of the operational global soil monitoring systems is operated by NOAA's CPC[6] (van den Dool, 2004). This system is based on the CPC's soil moisture model. This monitoring system provides the estimate of soil moisture percentile globally at about 2-month lag time. Along with percentile of soil moisture, it also provides anomaly and simulated value of soil moisture for the recent past (2 months ago), and for all months in the last year (Fig. 5.5).

[6] http://www.cpc.ncep.noaa.gov/products/Soilmst_Monitoring/gl_Soil-Moisture-Monthly.php

5.3.4 The Famine Early Warning Systems Network land data assimilation system

The Famine Early Warning Systems Network (FEWS NET) Land Data Assimilation System (FLDAS)[7] is another quasiglobal (50S−50N drought monitoring system. Recently extended to a global domain, the FLDAS supports the FEWS NET Food Security Outlook process. A custom instance of the NASA Land Information System (LIS) (Kumar et al., 2006) has been modified to work with the models and data commonly used by FEWS NET. The LIS contains the Noah (Ek et al., 2003) and VIC models (Liang et al., 1994). Comparisons of FLDAS outputs with independent verification data (satellite vegetation and soil moisture, observed streamflow) indicate good performance (McNally et al., 2016, 2017), and (Jung et al., 2017), and FLDAS outputs can be related to agricultural models to provide ag-impact models (Agutu et al., 2017; McNally et al., 2015) (Fig. 5.6).

5.4 Limitations of drought monitoring using land surface models

As indicated in this chapter, LSMs have become valuable tools for providing hydrologic and drought monitoring at national, continental, and global scales. The primary virtue of LSMs is that they provide estimates of hydrologic variables (such as soil moisture, runoff, and evapotranspiration) which, at a regional or global scales are rarely observed but are still very useful for estimating

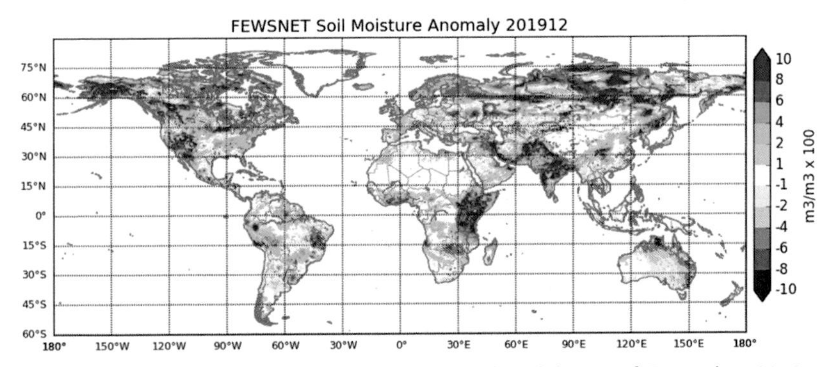

Figure 5.6 Monthly soil moisture anomaly over the globe as of December 2019, as displayed by the FLDAS. *FLDAS*, FEWS NET Land Data Assimilation System.

[7] https://ldas.gsfc.nasa.gov/fldas

drought severity and, hence, for drought management and the mitigation of socioeconomic losses due to drought. Nevertheless, this approach does have important limitations, as well as room for further improvement. The main limitations of this approach are as follows:

- The lack of high-quality in situ observations of atmospheric forcings, especially over developing and remote locations (e.g., high-elevation locations).
- The use of deterministic atmospheric forcings, potentially resulting in biased estimates of hydrologic variables.
- The lack of hydrologic observations to calibrate LSMs, mainly in developing regions.
- The use of deterministic parameters by the LSMs.
- A lack of representation of dynamical changes in vegetation.
- A lack of integration with models that simulate nonprecipitation water inputs such as reservoir models, groundwater models, and irrigation schemes.
- Basic uncertainties surrounding soil attributes, water holding capacity, and root depth.

5.5 Summary

The LSMs have been valuable tools in monitoring drought at national, continental, and global scales. This chapter provides an overview of the use of LSMs in hydrologic and drought monitoring. This chapter first provides a brief history of LSMs, and an overview of typical LSMs in terms of model physics and parameterization. Following which, this chapter provides an overview of three LSM-based hydrologic/drought monitoring systems that are currently in operation. Those systems include U.S.-focused NLDAS drought monitoring system, Africa-focused AFDM, and the CPC's global soil moisture monitoring system, and the NASA and FEWS NET's FLDAS. Finally, this chapter provides a quick overview of important limitations that currently exist based on the approach of using LSMs for hydrologic and drought monitoring.

References

Agutu, N., Awange, J., Zerihun, A., Ndehedehe, C., Kuhn, M., Fukuda, Y., 2017. Assessing multi-satellite remote sensing, reanalysis, and land surface models' products in characterizing agricultural drought in East Africa. Remote Sens. Environ. 194, 287–302.

Huffman, G.J., Bolvin, D.T., Nelkin, E.J., Wolff, D.B., 2006. The TRMM multisatellite precipitation analysis (TMPA): quasi-global, multiyear, combined-sensor precipitation estimates at fine scales [No title] J. Hydrometeorol. 8, 38–55.
Hamill, T.M., Bates, G.T., Whitaker, J.S., Murray, D.R., Fiorino, M., Galarneau, T.J., et al., 2013. NOAA's second-generation global medium-range ensemble reforecast dataset. Bull. Am. Meteorol. Soc. 94, 1553–1565.
Jung, H.C., et al., 2017. Upper Blue Nile basin water budget from a multi-model perspective. J. Hydrol. 555, 535–546.
Kumar, S.V., Peters-Lidard, C.D., Tian, Y., Geiger, J., Houser, P.R., Olden, S., et al., 2006. LIS—an interoperable framework for high resolution land surface modeling. Environ. Model Softw. 21, 1402–1415.
Liang, X., Lettenmaier, D.P., Wood, E.F., Burges, S.J., 1994. A simple hydrologically based model of land surface water and energy fluxes for general circulation models. J. Geophys. Res. 99, 14415.
McNally, A., et al., 2015. Calculating crop water requirement satisfaction in the West Africa Sahel with remotely sensed soil moisture. J. Hydrometeorol. 16, 295–305.
McNally, A., Shukla, S., Arsenault, K.R., Wang, S., Peters-Lidard, C.D., Verdin, J.P., 2016. Evaluating ESA CCI soil moisture in East Africa. Int. J. Appl. Earth Obs. Geoinf. 48, 96–109.
McNally, A., et al., 2017. A land data assimilation system for sub-Saharan Africa food and water security applications. Sci. Data 4, 170012.
Mitchell, K.E., Lohmann, D., Houser, P.R., Wood, E.F., Schaake, J.C., Robock, A., et al., 2004. The multi-institution North American Land Data Assimilation System (NLDAS): utilizing multiple GCIP products and partners in a continental distributed hydrological modeling system. J. Geophys. Res. 109. Available from: https://doi.org/10.1029/2003jd003823.
Nijssen, B., et al., 2014. A prototype global drought information system based on multiple land surface models. J. Hydrometeorol. 15, 1661–1676.
Sheffield, J., Goteti, G., Wood, E.F., 2006. Development of a 50-year high-resolution global dataset of meteorological forcings for land surface modeling. J. Clim. 19, 3088–3111.
Sheffield, J., et al., 2014. A drought monitoring and forecasting system for Sub-Sahara African water resources and food security. Bull. Am. Meteorol. Soc. 95, 861–882.
van den Dool, F.Y.A.H., 2004. Climate Prediction Center global monthly soil moisture data set at 0.5° resolution for 1948 to present. J. Geophys. Res. 109. Available from: https://doi.org/10.1029/2003jd004345.
Wang, A., Bohn, T.J., Mahanama, S.P., Koster, R.D., Lettenmaier, D.P., 2009. Multimodel ensemble reconstruction of drought over the Continental United States. J. Clim. 22, 2694–2712.

CHAPTER 6

Tools of the trade 3—mapping exposure and vulnerability

6.1 Exposure and vulnerability

This chapter builds on two excellent sources of international guidance on drought management and disaster risk reduction (DRR) resources provided by the Integrated Drought Management Program (IDMP, droughtmanagement.info) and the United Nations Office for DRR (UN DRR) (unisdr.org/). Our overarching goal is to describe, in general terms, how drought monitoring and prediction systems fit within social environments that reduce and respond to the impacts associated with droughts. This fit is an essential aspect of providing actionable information. While a detailed discussion of drought management and DRR is beyond the scope of this book, drought early warning systems (DEWS) clearly play a critical role in drought management and DRR. Understanding the basic principles of drought management and DRR can help one to inform more effective DEWS.

For guidance on drought management, readers should consult the resources provided by the IDMP (http://www.droughtmanagement.info). Wilhite (2012) is a seminal figure in this field, and his edited volume of articles is a valuable collection. The more recent *Drought: Science and Policy* (Iglesias et al., 2019) provides coverage of drought hazards as well as vulnerability, risk and policy, and drought-management experiences.

Each year, water insecurity costs the global economy more than 500 billion U.S. dollars.[1] To improve responses to this water insecurity, the World Meteorological Organization and the Global Water Partnership launched the IDMP to address drought issues more effectively. IDMP provides advice and guidelines to communities, countries, and regions affected by drought. The IDMP website provides valuable resources for those interested in learning more about systematic drought management.

One core concept promoted by the IDMP centers on the three pillars of drought management: monitoring and early warning; vulnerability and

[1] https://www.gwp.org/en/learn/KNOWLEDGE_RESOURCES/Global_Resources/securing-water-sustaining-growth/

Drought Early Warning and Forecasting
DOI: https://doi.org/10.1016/B978-0-12-814011-6.00006-3

impact assessment; and mitigation, preparedness, and response. These three pillars provide the building blocks of a successful drought policy. DEWS provide the foundation for effective, proactive drought policies. Vulnerability and impact assessments determine the primary impacts associated with droughts and their root causes. The UN DRR definition of vulnerability is *the conditions determined by physical, social, economic and environmental factors or processes which increase the susceptibility of an individual, a community, assets or systems to the impacts of hazards.*

According to the IDMP, the third pillar of drought management is drought mitigation, preparedness, and response. These measures and actions seek to reduce drought risk by reducing drought impacts and drought vulnerability through risk reduction and the identification of appropriate triggers to phase in and phase out mitigation actions. Mitigation actions minimize the impacts of hazardous events. Such actions are usually short-term responses that can arise during the onset or termination of a drought.

Short-term drought mitigation can be contrasted with longer term efforts to build coping capacity. Drought-coping capacity includes the combined attributes and resources within a society or community to manage and reduce drought risk and subsequently increase resilience. Resilience is the ability of a society or community to resist, absorb, and recover from the effects of a hazard in a timely manner. Coping capacity development is the process by which societies and communities systematically increase their abilities over time to achieve social and economic goals while decreasing vulnerability and increasing resilience. Drought-coping capacity development includes training, as well as continuous efforts to develop institutions, effective governance frameworks, and improved DEWS.

This chapter will focus on a deeper understanding of vulnerability and exposure, because these factors are most intimately linked to the development of effective DEWS. Ultimately, DEWS focus on the early identification of drought risks. This identification can then trigger mitigation or response actions. Risks occur within a multidimensional risk framework (risk = shock × vulnerability × exposure), so when assessing drought risks, we have to take vulnerability and exposure into account (Fig. 6.1).

Given the specific nature of drought, it is useful to expand this definition, replacing "shock" with "substantial water deficit." In practice, these water deficits will almost always have to do with the environmental, economic, agronomic, and societal setting, which will influence the level of exposure and vulnerability. These factors will also vary according to the various sectoral aspects of the droughts being considered, that is, agricultural droughts

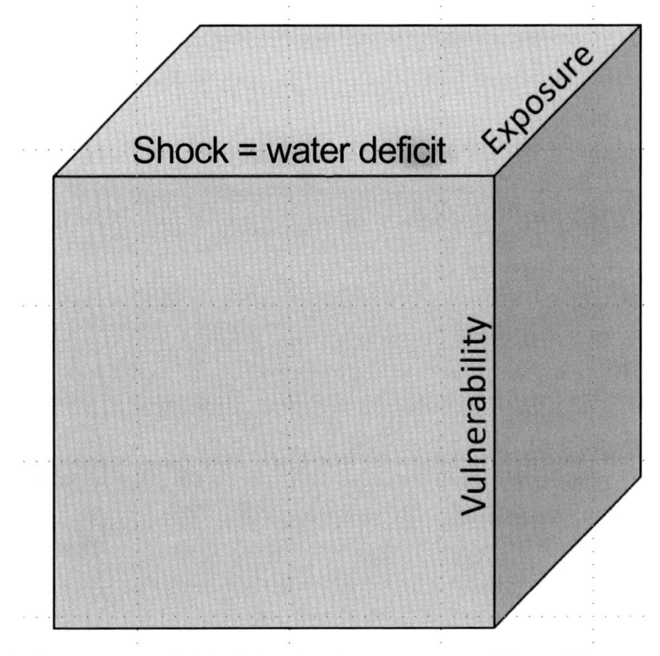

Figure 6.1 Components of risk: risk = shock × exposure × vulnerability.

may be different than rangeland/ranching/pastoral droughts, which may be different than regional water supply/hydropower droughts.

Exposure is a measure of who or what might be in harm's way. It is often tightly related to human population and economic activity. Measures of exposure can include the number of people or types of assets in an area. A rainfall deficit in an unpopulated region of a tropical rainforest or dry desert might not expose a large number of farmers to agricultural drought risk. On the other hand, there might be substantial herds of cattle, goats, or sheep in relatively unpopulated regions. Interactive tools that allow for the examination of environmental drivers such as precipitation deficits alongside exposure-related variables such as population can facilitate the identification of high-risk areas.

While there can be many types of exposure—and a complete list is beyond the scope of this book—agricultural and pastoral-ranching exposures are relatively straightforward. Exposure for these categories will be related to the location, quantity, and value of crops or livestock. The losses associated with poor harvests may be both direct and indirect. Direct impacts will relate to the immediate loss associated with reduced crop production. These can manifest as reductions in available food, for households dependent on local crop production, as well as reductions in farm-incomes. Secondary losses can also arise and may relate to increases in regional food prices. Many farming households both produce and consume food. Secondary losses can often arrive with a greater time lag and can include degradation of farm quality, drought-coping capacity, and resilience. Secondary drought impacts can extend beyond the farming household, reducing demand for regional farm labor, and reducing economic activity in related businesses.

Pastoral and ranching exposure is related to quantity and value of herds of animals. Reduced forage and water availability, combined sometimes with the direct effects of high temperatures, can reduce the health and weight of herd animals, decreasing their value. Ranchers and pastoralists may frequently need to purchase fodder, and sometimes even purchase water, in order to overcome the effects of drought on rangeland biomass and land surface runoff and water storage. In some regions such as sub-Saharan Africa or Mongolia, itinerant or semiitinerant pastoral communities constitute some of the most food insecure populations in the world. In addition to economic losses, these populations may also rely on herd animals as a direct source of calories from milk and meat.

Among poor food insecure pastoral and agricultural populations, the relative magnitude of exposure may be greater for pastoral communities, especially when losses are evaluated over a longer time span. In many cases, but not all, agricultural droughts impact a single, specific growing season. The farm household might experience catastrophic losses, but these have a limit—such as the entire potential harvest. The stakes for pastoral households may be greater. For most pastoral households, and many ranchers, herds represent very large accumulations of wealth. Often, these herds simultaneously act as economic reserves and as a source of income, and households try to increase the size of these herds over time. When droughts destroy these herds, years of economic investment can be lost in a matter of months. Years of better and wetter conditions will typically be required to recover from these losses.

Water supply for irrigated agriculture and consumption by industry, homes, and hydropower represent a third major and complex category of drought risk exposure. These potential impacts can be both direct and indirect and may involve water shortages a long way from the origin of a drought, as rivers connect water consumers with distant drought-stricken watersheds. One important aspect of these impacts can involve the expenses associated with the movement of water. Water is heavy, and the costs associated with pumping and transporting water from a water surplus to a water deficit region can be very high. For many countries, hydro-power is a major source of energy and income. Droughts in these countries can have substantial direct impacts associated with reduced power production, and secondary impacts associated with increased energy costs and/or limitations in the amount of energy available to support businesses and other societal activities.

In general, it is important to recognize that increases in population and economic activity tend to cause exposure to increase over time. This can increase drought risk without the actual frequency of droughts increasing. This can make it very hard to untangle the explicit contribution of low-frequency hydrologic changes, which may be associated with climate change. Fig. 6.2, for example, shows a time series of the number of global drought-related losses, both insured losses and total losses, based on data from the Munich Re catastrophe database. Munich Re is a global reinsurance company. They provide insurance to insurance companies, and, therefore, try to closely track global and national trends in catastrophic losses. In Fig. 6.2, we can see a large increase in the magnitude of annual climate catastrophes. Annual total losses hovered around 15 billion dollars in the early 1990s. In 2017 and 2018, total losses were around 34 and

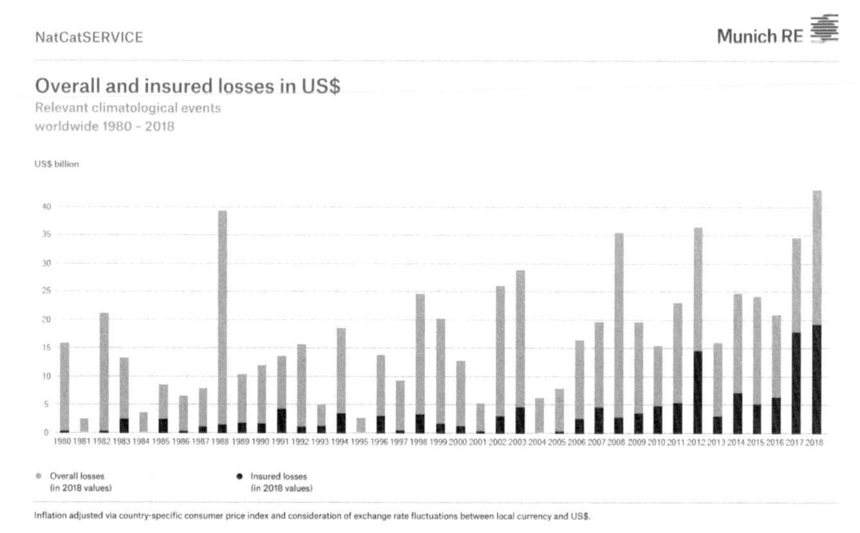

Figure 6.2 Time series of total and insured losses from Munich Re natural disasters database (https://natcatservice.munichre.com), accessed on August 13, 2019. Data shown in 2018 U.S. dollars, adjusted for inflation and exchange rate fluctuations. Data shown for climate disaster losses, which are primarily driven by droughts.

42 billion dollars. The United Nations believes that these droughts were related to increased global food insecurity and climate change (UN, 2018), but increases in human population and economic activity are also definitely contributing to these spikes in disaster frequencies.

6.1.1 Exposure—an example for East Africa

Exposure and water availability metrics can be combined to provide meaningful drought indices (see further discussion in Chapter 9: Sources of Drought Early Warning Skill, Staged Prediction Systems, and an Example for Somalia), and we provide here an illustrative example based on per capita water availability (PCWA) for East Africa (Harrison et al., 2018). This work builds off hydrologic simulations routinely produced by NASA scientists[2] using climate hazards center infraRed precipitation with stations (CHIRPS) data and the Famine Early Warning Systems Networks Land Data Assimilation System (FLDAS) (McNally et al., 2017). The FLDAS monitoring system also routinely produces maps (McNally et al., 2019) indicating per capita water stressed areas, based on the Falkenmark index (Falkenmark, 1989). This relatively simple index categorizes watersheds based on the amount of per capita runoff. Based on the total amount of per capita runoff, hydrologic basins can be categorized as facing absolute scarcity, scarcity, stress, or no stress.

While Harrison et al. (2018) found that most basins in East Africa have been experiencing modest *increases* in runoff, East African populations are increasing very rapidly (Fig. 6.3). For example, the population of Ethiopia has changed from about 38 million in 1980 to more than 100 million in 2017.

There is an important spatial component to these population increases. East Africa exhibits extreme variability in terms of its mean climate. Conditions range from extremely arid desert regions to moist tropical forests. When examining East Africa, one tends to find a covariation in mean precipitation and population density. In general, higher areas tend to receive more rain and support more people. The areas surrounding Lake Victoria are also humid and heavily populated. For areas with high fertility rates, high population densities are associated with high birth rates. Densely populated areas tend to grow much faster, in an absolute sense, than areas with relatively low population densities. Fig. 6.4 (Harrison et al., 2018) shows basin-level 2000–17 population growth in East Africa. In absolute terms, growth has been greatest in the Lake Victoria basin, with population increasing by more

[2] https://ldas.gsfc.nasa.gov/fldas

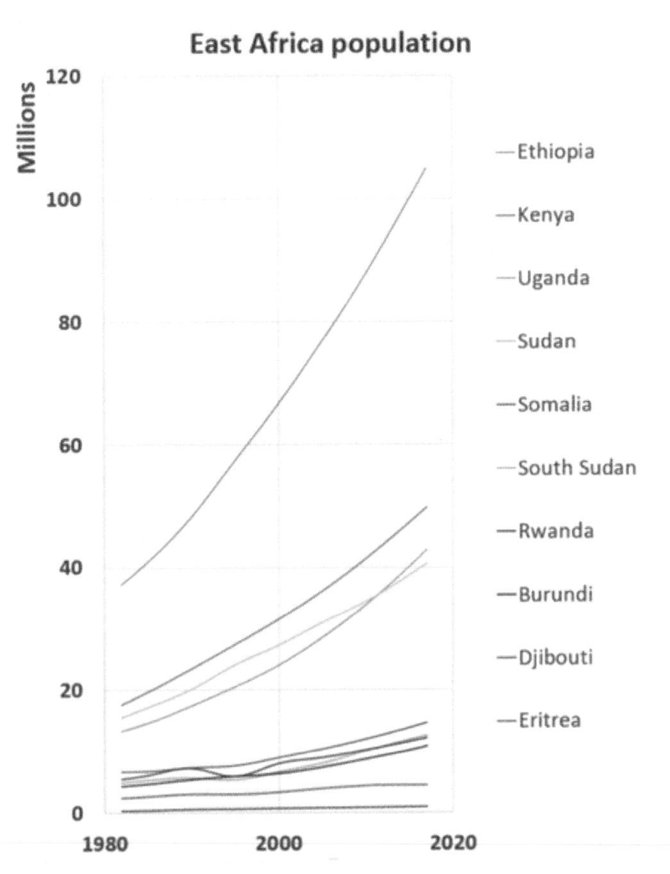

Figure 6.3 Time series of national United Nations population totals for East African countries. *Courtesy: Harrison, L., McNally, A., Shukla, S., Pricope, N.G., Funk, C.C., Galu, G., et al., 2018. Recent Water Availability Trends and Mid-21st Century Projections in East Africa. AGU Fall Meeting Abstracts.*

than 7 million between 2010–17 and 2000–09. Large (+5–7 million) changes are also observed in north-central Tanzania, central-western Kenya, and south-central Ethiopia. Rwanda, Burundi, north-western Tanzania, and north-eastern Ethiopia experienced basin-level population increases ranging between +3 and +5 million.

Runoff and population can be combined to map water stress risk using average PCWA (Harrison et al., 2018) characterized by Falkenmark category (Fig. 6.5). Blue areas are relatively water secure; dark-red regions face chronic severe water insecurity. The upper-left and upper-right panels in Fig. 6.5 depict, respectively, PCWA estimates for 1982–99 and 2010–17.

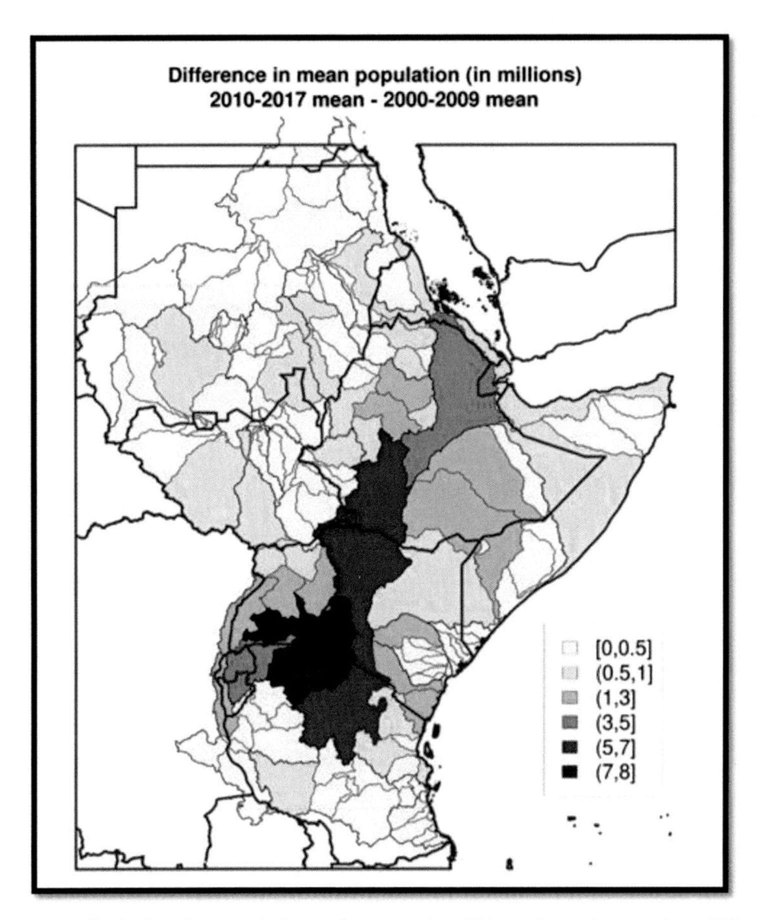

Figure 6.4 Basin-level population changes in Africa, based on the difference between 2010–17 and 2000–09 Worldpop data (http://www.worldpop.org.uk/data/). *Courtesy: Harrison, L., McNally, A., Shukla, S., Pricope, N.G., Funk, C.C., Galu, G., et al., 2018. Recent Water Availability Trends and Mid-21st Century Projections in East Africa. AGU Fall Meeting Abstracts.*

The bottom-left and bottom-right panels show projected water stress. These projections assume stationary runoff conditions based on observed 2010–17 values, while population values are assumed to increase, based on the average population projections provided by Boke-Olén et al. (2017).

For the historical PCWA maps, it is interesting to note that water insecure regions arise in both humid areas with high populations and in arid regions. High populations in the areas surrounding Lake Victoria result in water scarcity, while arid regions in eastern East Africa, Eritrea, and Sudan

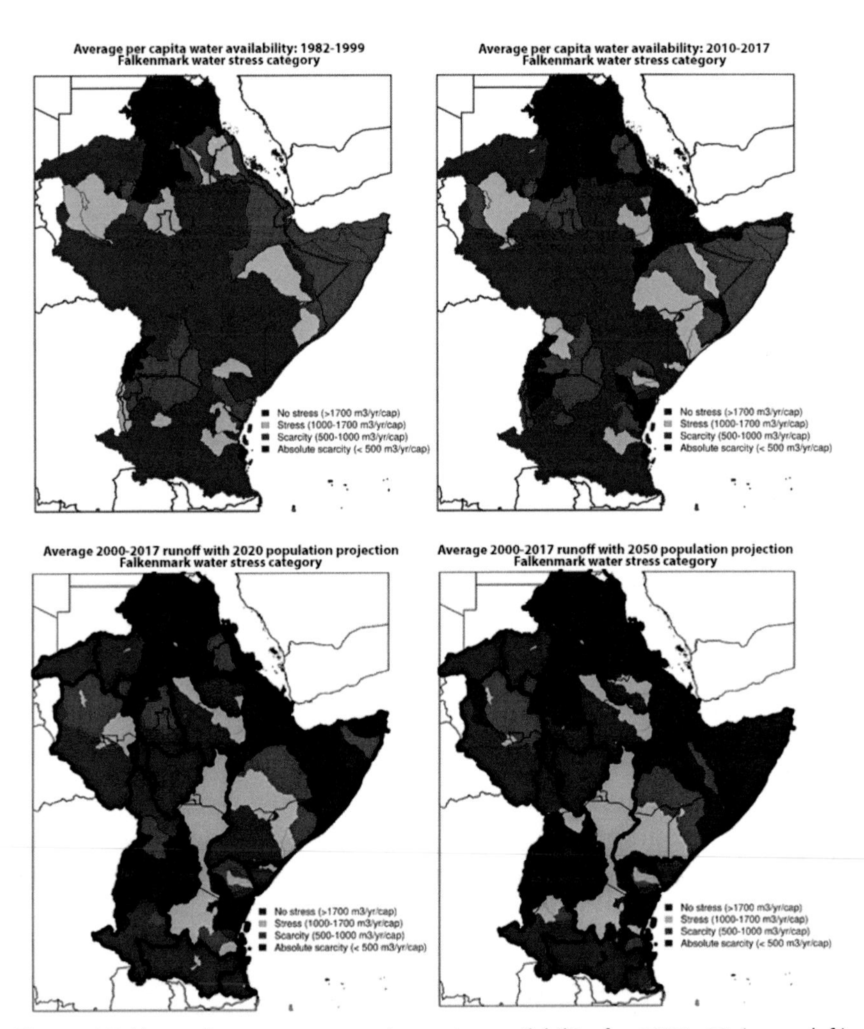

Figure 6.5 Maps of average per capita water availability for 1982–99 (upper-left) and 2010–17 (upper-right) along with projected water availability for 2020 (bottom-left) and 2050 (bottom-right). Per capita water availability characterized by Falkenmark water stress category. *Courtesy: Harrison, L., McNally, A., Shukla, S., Pricope, N.G., Funk, C.C., Galu, G., et al., 2018. Recent Water Availability Trends and Mid-21st Century Projections in East Africa. AGU Fall Meeting Abstracts.*

also exhibit substantial water stress. By 2020, these results indicate substantial expansions of water insecurity. Rwanda, Burundi, western Kenya, and most of Uganda are expected to experience chronic absolute water scarcity. Regions near the Pangani River and the Indian Ocean, near the border of Kenya and Tanzania, may also soon exhibit chronic absolute scarcity. Most of eastern Ethiopia, Somalia, Eritrea, and major portions of Sudan may experience severe water scarcity. Results for 2050 present further expansion, but these estimates emphasize that East Africa may be experiencing, in the near future, substantial increases in water stress due to increases in population-driven exposure.

6.1.2 Vulnerability

Vulnerability is the third major dimension of drought risk. The UN DRR defines vulnerability as *The conditions determined by physical, social, economic and environmental factors or processes which increase the susceptibility of an individual, a community, assets or systems to the impacts of hazards.* The same water deficit may arise in a region with the same number of people but have very different impacts based on the underlying physical, socioeconomic, and environmental factors. For example, a rancher or pastoralist who depends on fodder from a low-lying grassland close to the subsurface water table may be less vulnerable to droughts than a rancher or pastoralist relying on grassland with sandy soils (limited water-holding capacity) or grassland on upland areas. Some soils and landscapes will be able to retain water more effectively. Atmospheric water demand is another major physical determinant of drought vulnerability. This topic is treated in detail in Chapter 7—Theory—Understanding Atmospheric Demand in a Warming World, and we will refer to the estimates of atmospheric water demand as reference evapotranspiration (RefET). For now, though, we can all relate to variations in RefET by our shared understanding that at different times and places, the atmosphere is more likely to be able to draw moisture up from the land surface. When relative humidity is very low, and the Sun is shining strongly, and the weather is windy, moisture can be easily evaporated, transpired, and advected away from the planetary boundary layer. This can reduce the amount of moisture in soils and ponds, exacerbating the impact of precipitation deficits. In general, but not always, variations in precipitation and RefET tend to be inversely related. RefET tends to be high in hot, sunny, dry regions with low relative humidity and low precipitation. The spatial structure and temporal characteristics of drought risk can follow different characteristics according to our location along the precipitation/RefET gradient.

Another extremely important aspect of vulnerability relates to a household's, or community's, or country's social and economic susceptibility to drought impacts. This very often relates to poverty and wealth, for at least two reasons. Wealthier households, communities, and countries will be more likely to have invested in coping capacity to deal with hydrologic shocks. Wealthier households, communities, states, and nations will also be better situated to absorb the fluctuations in material and economic outputs associated with drought. For example, consider California, a very wealthy state that experiences relatively dry conditions. California has invested heavily in water infrastructure, with extensive water-storage facilities spread throughout the state. These large investments make California less susceptible to interannual fluctuations in water supply. Furthermore, individual households, communities, and counties within California tend to be relatively wealthy, and this wealth helps them limit the impacts of water shortages.

Conversely, one might consider a smallholder farm household in a poor arid country. Interannual water, food, and income storage options are often very low. These households eke out a difficult living from the soil, with a very limited capacity to build up economic reserves or store water or food from previous rains and harvests. These poor farm households may spend 60%–70% of their household income on food and may be highly reliant on income from farming. When crops fail, less food can be garnered from fields and gardens, incomes from crop sales are diminished, and local food prices often spike, creating a threefold path to food stress.

6.1.3 An Ethiopia case study

We next present a more detailed discussion of drought exposure and vulnerability in one food insecure African country: Ethiopia. This discussion uses materials from the 2016 Atlas of Ethiopian livelihoods, produced by the Food Economy Group and government of Ethiopia.[3] With more than 112 million inhabitants in 2019, and an area of some 1.04 million square kilometers, Ethiopia provides a study in contrasts. Ethiopia has many contrasting rainfall regimes (Fig. 6.6) that are tightly coupled to the country's variations in elevation (Fig. 6.7) and population. In general, mean precipitation tends to increase from east to west, and at higher elevations. The arid low-land regions in the east (tan in Fig. 6.6) are too dry to support agriculture. People in these regions rely on pastoral livelihoods. The high-elevation central highlands extend over 2000 m and are highly populated.

One measure related to both exposure and vulnerability is the average size of Ethiopian farms, shown in Fig. 6.8 in units of "hectares" per household. The green and blue areas denote the regions in Ethiopia with substantial agricultural exposure. These are regions where households tend to

[3] http://foodeconomy.com/wp-content/uploads/2016/02/Atlas-Final-Web-Version-6_14.pdf

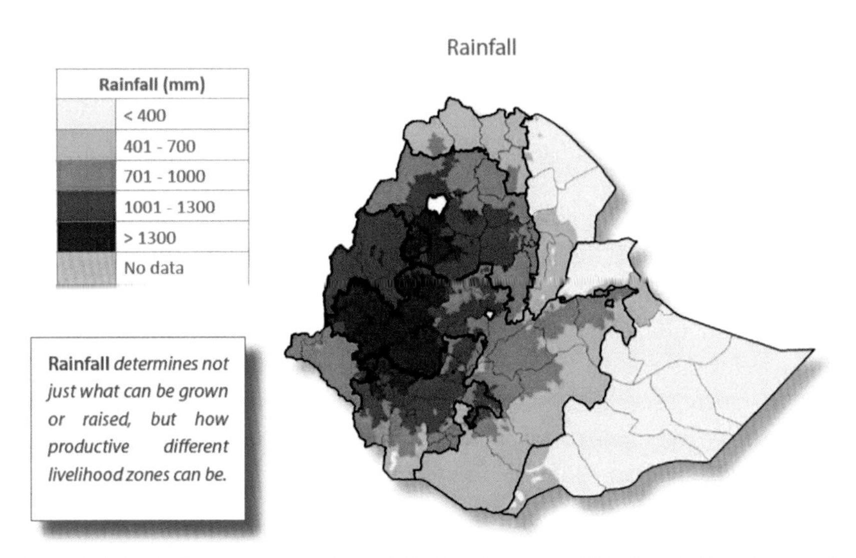

Figure 6.6 Ethiopian mean annual precipitation. *Courtesy: Ethiopian livelihood analysis.*

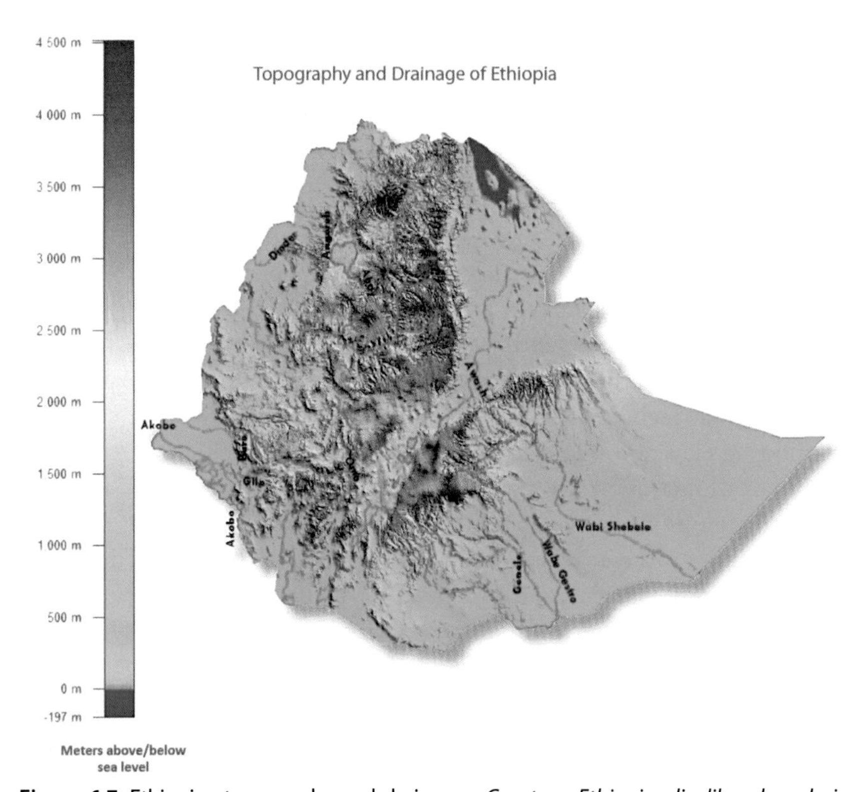

Figure 6.7 Ethiopian topography and drainages. *Courtesy: Ethiopian livelihood analysis.*

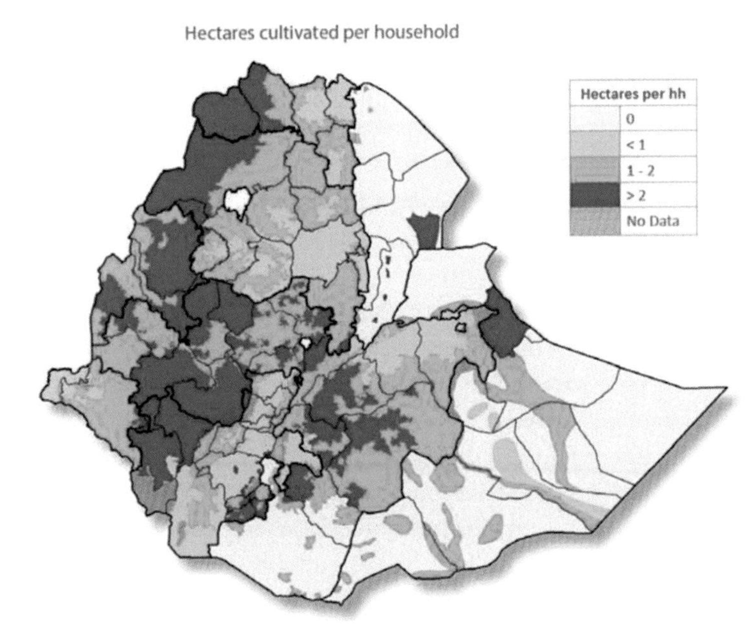

Figure 6.8 Ethiopian average-household farm size in hectares. *Courtesy: Ethiopian livelihood analysis.*

depend on agriculture. One hectare is 100 m^2 or 2.5 ac. A 1- or 2-ha farm is, therefore, quite small. This small farm size, combined with endemically low crop yields, results in relatively small levels of per-household crop production. The values of these crops can be further limited by a lack of market access—due to limited, expensive transportation and storage opportunities. African smallholder farmers, therefore, often exhibit substantial vulnerability due, at least in part, to their limited economic opportunities, and limited employment opportunities. Agricultural shocks due to water shortages can impact these poor households through multiple paths. Droughts can reduce household incomes from crop sales, and the local household food consumption of local farm produce. Almost all poor households also purchase food, often spending 60%–70% of limited household incomes on food purchases. Droughts can decrease household incomes and increase food prices, creating a triple threat of decreased incomes, increased external food costs, and decreased on-farm food availability.

Vulnerable populations tend to be poor, and they experience severe food insecurity when they are unable to purchase, or access, adequate food (Sen, 1981). The poorest must compete for limited resources with the more well-to-do members of the population. Ironically, increasing wealth by the middle and upper class may be contributing to increased food stress in some developing nations. For example, consider World Bank estimates of per capita incomes for average and the poorest 20% of Kenyans and Ethiopians (Fig. 6.9).

These estimates are based on World Bank Development indicator statistics and have been produced by combining gross national income values, based on the Atlas method and expressed in current U.S. dollars, population, and estimates of the income share held by the lowest 20%. These latter factors are only observed intermittently, and linear interpolation has been used to produce continuous estimates. Ethiopia and Kenya, however, have recent surveys from 2015, during which the shares of income estimates for Ethiopia and Kenya were 6.6% and 6.2%, respectively. Just considering these national scale indicators, one might anticipate a dramatic decline in vulnerability to food insecurity, since average incomes for poor households have increased by more than 300% since the early 1990s. These national averages, however, obscure important fluctuations in subnational household-level incomes, as well as the potential implications of fluctuations in prices.

We can explore these income differences by plotting (Fig. 6.10) the difference between the national average incomes and the annual incomes

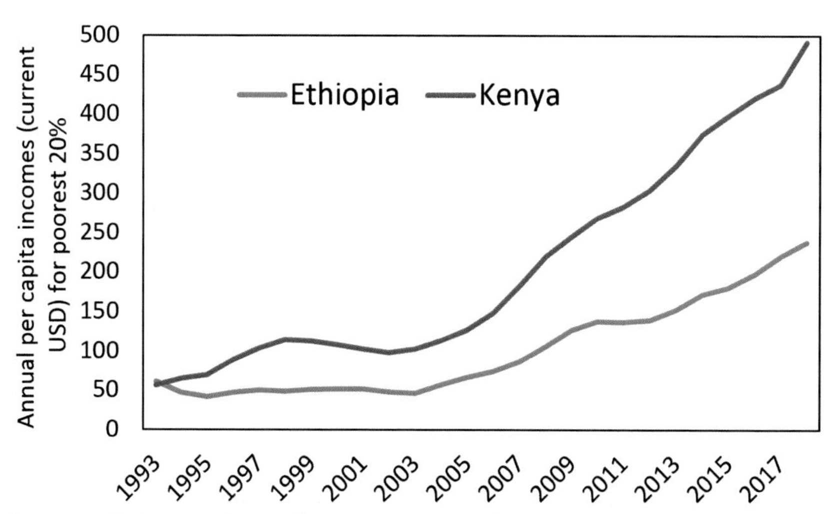

Figure 6.9 Estimates of annual per capita income for the poorest 20% of the population. *Courtesy: Chris Funk.*

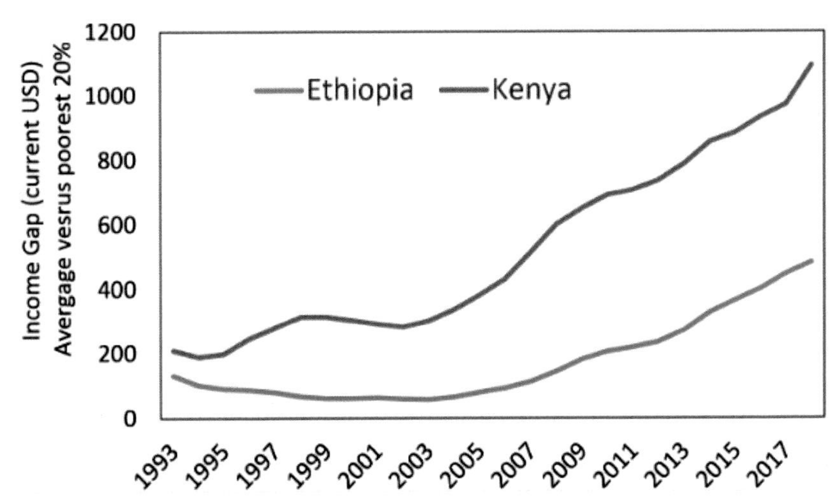

Figure 6.10 Estimates of annual "income gap" based on the difference between mean per capita incomes, and incomes for the poorest 20% of the population. *Courtesy: Chris Funk.*

for the poorest 20%. The wealthier 80% of the populations in Ethiopia and Kenya (and almost all nations) have had their incomes increase substantially more than the poorest 20%. In Kenya, this gap has increased by approximately 800 dollars annually. In Ethiopia the change has been about 300 dollars annually. So, while the poorest Kenyans and Ethiopians have been, on average, wealthier, they have also fallen farther behind economically in terms of their purchasing power.

These income gaps may be interacting with large fluctuations in commodity and food prices. Commodity prices in Ethiopia and Kenya remain very volatile, as illustrated by Fig. 6.11, that shows nominal FAO GIEWS wholesale maize prices. These values have not been adjusted for inflation and hence may overemphasize recent price increases. On the other hand, incomes for the poorest households may not be keeping pace with national inflation rates, as implied by Figs. 6.9 and 6.10. Ethiopian maize prices spiked in September–October of 2017 and have remained relatively high since. Kenyan maize prices exhibit a clear periodicity that aligns with major recent droughts in 2010/11, 2016/17, and 2019. During these Kenyan price spikes, we see a doubling of maize prices. For poor households living on a dollar or two a day and spending 60%–70% of their household income on food, such spikes can lead to dramatic food access limitations.

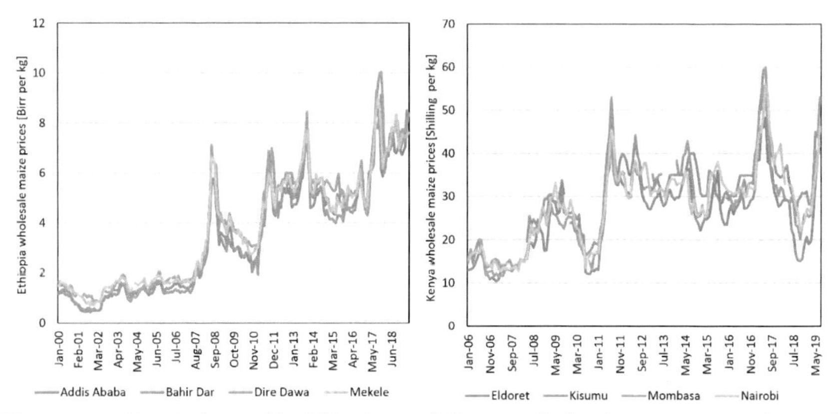

Figure 6.11 Nominal monthly Ethiopian and Kenyan wholesale prices. Data obtained from the United Nations Food and Agriculture Organization Global Information and Early Warning System. http://www.fao.org/giews/food-prices/tool/public/#/dataset/domestic. *Courtesy: Chris Funk.*

6.2 Conclusion

While a full treatment of drought management and DRR is beyond the scope of this book, understanding how exposure and vulnerability can act to magnify the impact of a drought shock (Fig. 6.1) is an important part of effective drought early warning. Depending on the context, the same precipitation deficit may have very different impacts. In a wet region with soils with deep water-holding capacities, rainfall reductions may actually enhance crop yields, because plant growth in such regions is often limited by the amount of available radiation, not water. Under such conditions, sunny skies during the growing season may enhance plant growth. Conversely, in regions with high RefET or porous soils, water deficits may have serious impacts. There is a temporal aspect to such changes in vulnerability, since plant growth and evapotranspiration in many humid and semihumid regions may be water-limited at the beginning and end of the growing season but are typically radiation-limited in the middle of the growing season.

We have also briefly explored, how, ironically, growing prosperity does not necessarily mean an end to drought risk. At a global scale, increasing population and economic activity appear related to substantial increases in drought exposure and associated economic losses (Fig. 6.2). While countries such as Ethiopia and Kenya are experiencing rapid economic growth, the poorest people in these countries are still desperately poor, with annual incomes of 150–450 U.S. dollars (Fig. 6.9). As the income gap between the poor, middle, and upper classes increases in these countries (Fig. 6.10), the vulnerability of these poor households to drought-related food price spikes (Fig. 6.11) may actually be increasing too. At a global scale a recent report by the United Nations Food and Agricultural Organization (UN, 2018) finds a recent increase in the global number of undernourished people and underscores the need to develop enhanced climate resilience, early warning systems, and DRR.

As demonstrated by our brief analysis of exposure impacts in East Africa (Harrison et al., 2018) (Figs. 6.3–6.5), we are likely to see increased water stress as human populations continue to grow. Even in a world without climate change, drought risks are likely to increase. Effective DEWS will play a critical role in managing and mitigating these risks.

References

Boke-Olén, N., Abdi, A.M., Hall, O., Lehsten, V., 2017. High-resolution African population projections from radiative forcing and socio-economic models, 2000 to 2100. Sci. Data 4, 160130.

Falkenmark, M., 1989. The massive water scarcity now threatening Africa: why isn't it being addressed? Ambio 112–118.

Harrison, L., McNally, A., Shukla, S., Pricope, N.G., Funk, C.C., Galu, G., et al., 2018. Recent Water Availability Trends and Mid-21st Century Projections in East Africa. AGU Fall Meeting Abstracts.

Iglesias, A., Assimacopoulos, D., Van Lanen, H., 2019. Drought: Science and Policy. Wiley-Blackwell, p. 251.

McNally, A., et al., 2017. A land data assimilation system for sub-Saharan Africa food and water security applications. Sci. Data 4, 170012.

McNally, A., et al., 2019. Hydrologic and agricultural earth observations and modeling for the water-food nexus. Front. Environ. Sci. 7.

Sen, A., 1981. Poverty and Famines: An Essay on Entitlement and Deprivation. Oxford University Press.

UN, 2018. The State of Food Security and Nutrition in the World 2018: Building Climate Resilience for Food Security and Nutrition. FAO.

Wilhite, D.A., 2012. Drought Assessment, Management, and Planning: Theory and Case Studies: Theory and Case Studies, vol. 2. Springer Science & Business Media.

CHAPTER 7

Theory—understanding atmospheric demand in a warming world

7.1 Background

This book has stressed how drought arises from the interplay of moisture supply and moisture demand. Moisture supply originates with precipitation, which is then transferred and stored by soils, streams, and reservoirs. Conservation of mass ensures that what comes down must go up via evaporation or transpiration, or down via deep percolation into the soil. The upward moisture flux from the land to atmosphere is commonly referred to as evapotranspiration (ET). The focus of this chapter—the upward limit of the ET moisture flux at any given place or time—is commonly referred to as reference ET (RefET), or in some instances, potential ET (PET).[1] In general, RefET refers to an estimate of the ET for a specific crop under well-watered conditions. The older PET term is not as specific. Hence, we will use PET in our discussion of the historic development process, and RefET to describe the modern applications. The "Ref" in RefET refers to a standardized reference crop, typically a short well-watered grass. While ET_0 is another very commonly used term in academic literature, we use RefET throughout this book because of its clear meaning.

Evaporation refers to the process by which liquid water in the soil or surface of plants changes phase to become gaseous water vapor and is vertically transported by turbulent atmospheric motions into the atmosphere. The conversion of liquid water to vapor (vaporization) requires both energy from the Sun and vertical atmospheric motions (convection) that extract the evaporated moisture from the lowest level of the atmosphere.

[1] This chapter's description benefits substantially from the excellent and globally used description provided by the United Nations Food and Agriculture Organization guidelines for calculating crop evapotranspiration. <http://www.fao.org/3/X0490E/x0490e00.htm>.

Drought Early Warning and Forecasting
DOI: https://doi.org/10.1016/B978-0-12-814011-6.00007-5

Without adequate insolation, there will not be adequate energy to support the vaporization of liquid water. Without adequate convection, the lowest atmospheric level will become saturated, and moisture will condense as rapidly as it evaporates. Hence, there must be less water vapor in the lower atmosphere than at the surface. Over time, however, in the absence of wind, atmospheric mixing will equilibrate the amount of moisture at these two levels. Low-level winds, therefore, are a third critical factor for evaporation. Shade and the amount of available soil moisture will also impact evaporation rates.

Transpiration refers to the process by which the liquid water in plants is vaporized and removed by the atmosphere. Most crops and many plants lose most of their water through small pores in their leaves called stomata. These stomata are used to draw carbon dioxide (CO_2) out of the atmosphere, which can be combined with water and transformed into sugars and carbohydrates via photosynthesis. This process is referred to as plant respiration. The same small stomatal openings that let in CO_2 also emit water vapor from the leaves, as it transpires into the atmosphere. Like evaporation, transpiration is also controlled by the local humidity gradient, air temperature, radiation, and wind. The rate of transpiration is further controlled by the physical characteristics of the plant. In particular, the total surface area covered by the leaves of many plant species typically changes across the varying stages of their phenological cycles, thus changing the total area over which transpiration can occur. In addition, the size of the stomatal openings, as well as the way in which they respond to changes in the plant's water status and environmental conditions, such as air temperature, also vary for different plant species.

In the mid-20th century the American geographer and climatologist Charles Warren Thornthwaite (1899–1963) incorporated such thinking in his development of climate classification systems. To represent atmospheric water demand, Thornthwaite introduced, in the 1940s, the concept of "PET," as the amount of water which could evaporate and transpire from a surface if unconstrained by water supply (Thornthwaite, 1948).

A central insight of Dr. Thornthwaite was that precipitation alone was insufficient to categorize climate: "We cannot tell whether a climate is moist or dry by knowing the precipitation alone. We must know whether precipitation is greater or less than the water needed for evaporation and transpiration .... Where precipitation is in excess of water need, the climate is moist. Where the water deficiency is large in comparison with the need, the climate is dry. Where precipitation and water need are equal or nearly equal, the climate is neither humid nor arid" (Thornthwaite, 1948). To address the question, Thornthwaite first considered the various factors contributing to both evaporation and transpiration. Taken together, these two moisture fluxes are described as ET.

In Thornthwaite's seminal 1948 paper, he argues that in many parts of the world, the amount of available water is limited, and resulting actual ET (AET) values that are less than that region and season's PET. At a set location on the Earth, most places also experience a seasonal cycle with a dry season (when RefET exceeds PET), and a wet season (when PET equals RefET). Thornthwaite (1948) lays out a number of physical and biological processes that link increases in temperature to increases in transpiration or evaporation. Plants use transpiration to dissipate heat, helping to optimize leaf temperatures. Plant growth rates are also related to increases in temperature. This logic, and a lot of empirical analysis, leads to the following Thornthwaite equation for PET, as laid out in a recent comparison of numerous PET/RefET formulations (Maes et al., 2019):

$$T_{\text{eff}} < 0 \text{then PET} = 0$$

$$0 < T_{\text{eff}} < 26 \text{then PET} = \alpha_{\text{Th}} \left(\frac{10 T_{\text{eff}}}{I} \right)^{b} \left(\frac{N}{360} \right)$$

$$26 < T_{\text{eff}} \text{then PET} = -c + d T_{\text{eff}} - e {T_{\text{eff}}}^{2}$$

where T_{eff} is the effective average daily temperature in degrees Celsius $[T_{\text{eff}} = 0.36(3T_{\max} - T_{\min})]$, $\alpha_{\text{th}} = 16$, if the daily temperature is less than 0. I is a function of the sum of the annual monthly average temperatures at each location $I = \left(\sum \overline{T}_{\text{month}}/5\right)^{1.514}$. N is the number of daylight hours, b is a parameter depending on I while c, d, and e are empirical constants (Maes et al., 2019).

While rather convoluted, Thornthwaite's formulation was relatively easy to calculate with available climate data—that is, monthly air temperature data. Fig. 7.1 shows Thornthwaite's estimates of RefET for the continental United States. This ease of calculation has led to the continued use of temperature-based RefET estimates. It should be noted, however, that such approaches often lead to a tendency to overestimate the influence of temperature (and global warming) on RefET. More

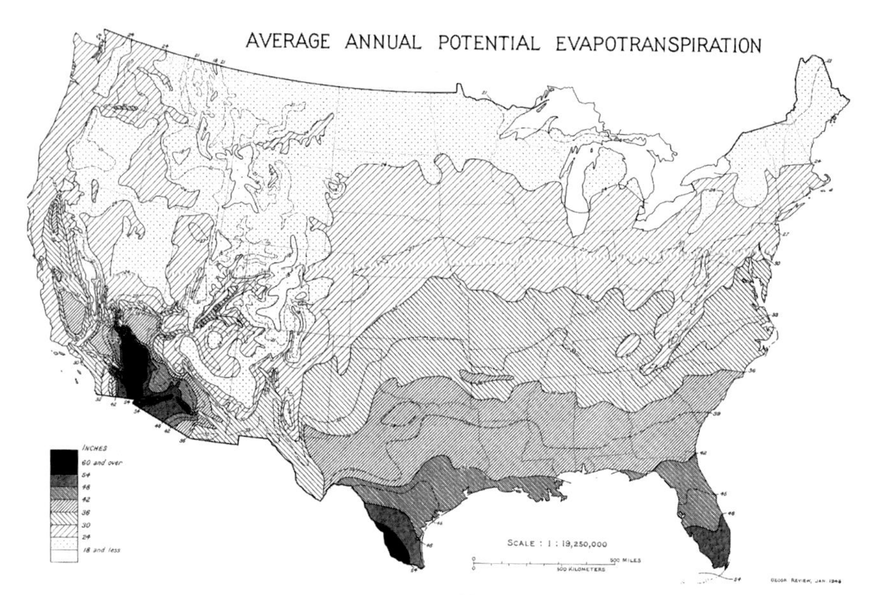

Figure 7.1 Average annual PET, as estimated by Thornthwaite in 1948. *PET*, Potential evapotranspiration.

physical models, such as the Penman–Monteith model, discussed later, are generally preferable since we no longer face the same data limitations.

It is quite interesting that just as Thornthwaite was publishing his largely empirical results, the physicist Howard Latimer Penman (1909–84) was attacking the problem from a theoretical perspective (Penman, 1948). This endeavor has led to the most widely used RefET formulation, the Penman–Monteith equation. It is worth reviewing Penman's (1948) formulation. Penman's particular genius was to frame PET as founded on two basic requirements: (1) a supply of energy that enables water molecules to transition from a liquid to a gaseous phase and (2) a mechanism for removing the resulting water vapor. Without "(1)" water will be unable to evaporate in the first place. Without "(2)" a water molecule may evaporate into a layer just above the land surface, but rising humidity levels will produce an equal and offsetting rate of condensation. Thus an evaporating land surface must provide a "sink" by which the moisture is transferred away from the land surface.

Penman (1948) treats the "sink" constraint first, since it was fairly well understood at the time, having been treated a century prior by the British physicist and chemist John Dalton (1766–1844). Dalton introduced the atomic theory into chemistry—a theory in which chemical elements comprise identical, indivisible atoms of different molecular weights. Dalton produced an expression for the rate of evaporation, which we present here as a function of water vapor pressure (e_a) and saturation water vapor pressure (e_s). The water vapor pressure is the atmospheric pressure (in Pascals) of the water vapor molecules in a parcel of air. The saturation vapor pressure is the upper limit of the water vapor pressure for a fixed air temperature and density. When e_a equals 0, the air will have a relative humidity of zero. When $e_s = e_a$ the air will be fully saturated, and the relative humidity will be 100%. Dalton's evaporation formula, as expressed by Penman, was

$$E = (e_s - e_a)f(u)$$

$f(u)$ in this formulation is a function of horizontal wind speed (u) and E is evaporation. While this term is based on horizontal wind speeds, it is really parameterizing (for the most part) vertical mixing in the lowest portion of the atmosphere. As air moves across the land surface, turbulent flows mix the lowest several meters. Faster winds produce more turbulent flow, more mixing, and a greater potential atmospheric "sink" for water vapor, except when $e_s = e_a$. When the air is fully saturated, the air being mixed down will be just as moist as the air being mixed up, and the drain on the atmospheric "sink" will be plugged. A very common expression for the $e_s - e_a$ term is vapor pressure deficit (VPD).

Penman's other major constraint on E was the total amount of energy available to transition water molecules from their liquid to gaseous phase. This total energy is based on the total amount of "net radiation" (R_n) minus the loss of heat energy into the soil (G). The net radiation is a combination of the incoming longwave and shortwave radiation, reflected outgoing shortwave radiation, and emitted outgoing longwave radiation. Shortwave radiation is radiation in the near infrared, visible or higher electromagnetic frequencies. Longwave radiation is radiation in the mid-infrared or lower frequencies. The R_n term is dominated by an approximate balance between downwelling shortwave radiation (i.e., sunlight) and upwelling longwave radiation emitted by the Earth's surface. In places where moisture can be evapo-transpired, the difference between these two terms is largely made up by the energy associated with the AET flux, that is, in approximate terms, the energy provided by sunlight $\approx$ emitted heat energy + energy used to vaporize water.

7.2 Reference evapotranspiration resistance terms

We now follow the classic description of crop RefET as laid out by the United Nations Food and Agricultural Organization's "Irrigation and Drainage Paper 56" (Allen et al., 1989). Penman's original formulation (1948) was derived from estimates based on evaporation from an open water surface using standard climatological weather observations. Later scientists came to recognize that Penman's original formulation did not consider two important sources of ET resistance from the land surface. The first resistance term, the "bulk surface resistance" (r_s), parameterizes the net resistance experienced by water vapor flowing from the soil surface, from leaves, or through stomata. The second resistance term, the "aerodynamic resistance" (r_a), describes the resistance associated with vertical moisture transports from air moving across and through complex vegetated surfaces. The term "resistance" here is analogous to resistance in an electrical current. In an electrical circuit, voltage (V) equals the product of current (I) and the resistance (R), $V = IR$. For a set voltage, increasing the resistance decreases the current. Current in this instance would be analogous to RefET. Resistances in series multiply, so for RefET, $R_{\text{total}} = r_s \times r_a$. Fig. 7.2 shows this relationship schematically (see footnote 1). John Lennox Monteith[2] contributed a great deal to the development of these resistance terms and used these terms to improve Penman's original formulation. The combined Penman–Monteith algorithm is described in the next section (Allen et al., 1989).

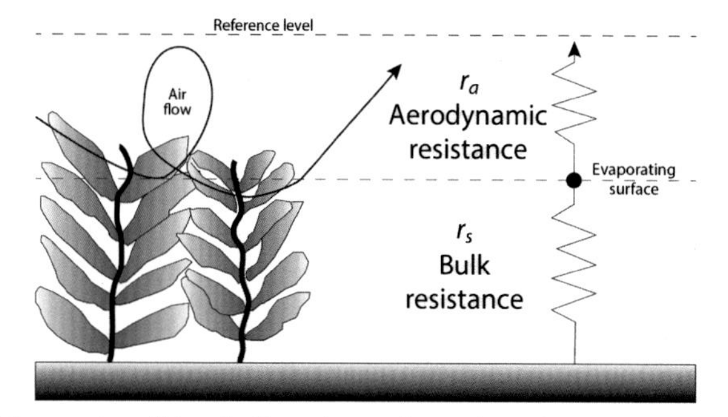

Figure 7.2 Schema describing bulk and aerodynamic resistances.

[2] https://royalsocietypublishing.org/doi/10.1098/rsbm.2014.0005

7.3 Defining reference crop evapotranspiration

Reference crop ET (i.e., RefET) is defined as the ET rate from a well-watered large expanse of a reference surface, which is almost always a grass reference crop with defined characteristics. The definition of RefET is typically superior to other denominations such as PET, because it is clear and well defined. It should be noted, however, that in applications such as ecosystem modeling and hydrologic modeling, a crop-centric definition of RefET might not be appropriate. Historically, the concept of RefET was introduced to support the study of evaporative demand independent of considerations of crop management and crop type (Allen et al., 1989). Importantly, the only factors influencing RefET are weather: radiation, humidity, temperature, and wind speeds. RefET estimates the evaporative demand of the atmosphere at a given location, at a given time of the year, without reference to soil parameters or crop specifics.

The FAO 56 reference crop is assumed to be a short (0.12-m high) green grass with a fixed surface resistance (r_s) of 70 s m^{-1} and an albedo of 0.23. This albedo means that the surface is quite dark and only reflects 23% of the downwelling shortwave radiation. The aerodynamic resistance (r_a) will be a function of wind speed: $r_a = 208/u_{2m}$, where u_{2m} is the wind speed at 2 m.

7.4 The FAO 56 Penman–Monteith formulation

We are now ready to present the Penman–Monteith equation. While the equation is quite complex, we will see that it can be understood as the combination of a radiative and ET term.

$$\text{RefET} = \frac{0.408\Delta(R_n - G) + \gamma\big(900/(T+273)\big)u_{2m}(e_s - e_a)}{\Delta + \gamma(1 + 0.34u_{2m})}$$

RefET in this equation is expressed in mm day^{-1}. $(R_n - G)$ denotes the energy provided by radiation (net radiation minus heat loss into the Earth), $u_{2m}(e_s - e_a)$ describes the potential vertical moisture flux produced by atmospheric mixing in the planetary boundary layer. This term is a function of 2m wind speed (u_{2m}) and the VPD ($e_s - e_a$).

This formulation makes use of the psychrometric constant (γ) and the slope of the saturation water vapor curve (Δ). It is important to

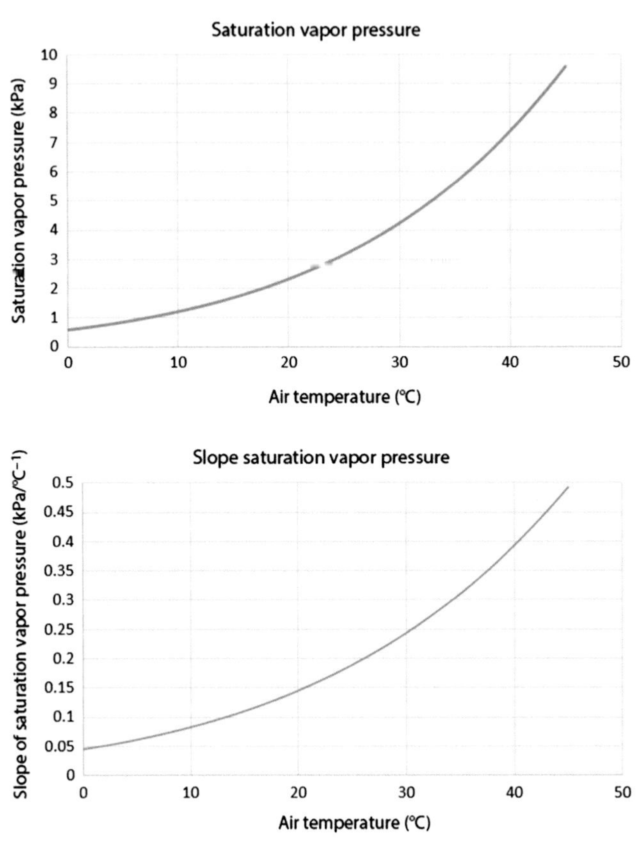

Figure 7.3 Plots of air temperature and saturation vapor pressure (kPa) (top) and air temperature and the temperature slope of saturation vapor pressure (bottom) (kPa °C^{-1}).

note that there is a nonlinear relationship between air temperature and saturation vapor pressure (e_s), as well as between air temperature and the saturation vapor pressure–air temperature slope (Fig. 7.3). Between freezing temperature (0°C) and very warm temperatures (45°C), we see a ~10-fold increase in the water-holding capacity of air. At saturation (100% humidity), very warm air can hold much more water than very cold air. This temperature dependency affects both the value of e_s and Δ (top and bottom plots in Fig. 7.3).

The FAO 56 formulation can be broken into two components: a radiation term and a vertical moisture transport term, both of which are modifications of the original formulation proposed by Penman in 1948 (Eq. 16 in Penman, 1948). We can rewrite the FAO 56 formulation to emphasize this decomposition.

$$\mathrm{RefET} = \frac{0.408\Delta(R_n - G)}{\Delta + \gamma(1 + 0.34u_{2m})} + \frac{\gamma\left(900/(T + 273)\right)u_{2m}(e_s - e_a)}{\Delta + \gamma(1 + 0.34u_{2m})}$$

Two basic factors control the RefET magnitude. There must be a supply of energy to provide the energy required to transform liquid water into water vapor ($R_n - G$), and there must be some mechanism for removing the vapor from the planetary boundary layer [$u_{2m}(e_s - e_a)$], that is, there must be a sink for vapor. The Δ and γ terms in the Penman–Monteith equation arise from the expected relationship between air temperature and vapor pressure gradients. The ($1 + 0.34u_{2m}$) term reflects the combined influence of bulk and aerodynamic resistance terms for the reference crop cover.

7.5 Temperature alone is insufficient to estimate reference evapotranspiration

Ultimately, there are three factors that tend to dominate RefET variations in space and time: changes in radiation, wind speed, and VPD. Changes in radiation and VPD often relate to changes in air temperature, since radiation increases temperatures, and there is a direct nonlinear relationship between e_s and air temperature (Fig. 7.3, top). The use of temperature-based estimation procedures, like that developed by Thornthwaite (1948) can be convenient but fail to represent the key processes driving RefET. Hence, such approaches can be misleading, especially when used to assess the impacts of climate change. Climate change can modify the terms in the FAO 56 formulation in complicated ways. We can expect longwave radiation to increase by a few watts per meter squared (W m^{-2}), but it may also be difficult to predict variations in cloud cover and shortwave radiation. Increasing air temperature will enhance saturation vapor pressures (e_s), but they will also increase the slope of saturation vapor pressures (Δ) (Fig. 7.3), which appears in the denominator of the FAO formulation. These complexities and considerations suggest caution when assessing the potential impacts of climate change on RefET.

7.6 Reference evapotranspiration decompositions and Morton's complementary hypothesis

While beyond the scope of this book, interested readers may wish to examine research focused on decompositions of RefET fields (Hobbins et al., 2012). This work analytically decomposes the RefET signal into forcing by radiation, wind speeds, and VPD variations. Another important avenue of research builds on Bouchet's description of the "complementary relationship" (Bouchet, 1963) between RefET and AET. In general, it is important to recognize that the hydrologic balance at any given location and time period on the planet will either be energy-limited or water-limited. Energy-limited areas have ample soil moisture and will have AET rates capped by the amount of available radiation. Water-limited areas, on the other hand, will have insufficient soil moisture and vegetation to support an AET flux equivalent to the net radiation balance. Under water-limited conditions, variability in AET drives a complementary variation in RefET through energetic exchanges across the land–atmosphere interface (Hobbins et al., 2016). In general, neglecting heat storage, the energy balance at a given location on land can be written as $(R_n - G) = \lambda \cdot + H$, where λ is the latent heat of vaporization, AET the amount of water vapor evaporated and transpired, and H is the sensible heat flux from the surface. H and $\lambda \cdot \text{AET}$ are expressed as energy fluxes in W m^{-2}. H represents the vertical advection of heat energy by atmospheric mixing in the planetary boundary layer. H typically covaries with $u_{2m}(T_s - T_{a2m})$. T_s and T_{a2m} represent surface and $2m$ air temperature, respectively.

When a location is water limited and no new water inputs occur, AET will gradually decrease, while H, temperature, e_s and VPD will increase. Following Hobbins et al. (2016) treatment, these relationships lead to a general complementary relationship, valid for water-limited conditions, which we can approximate as:

$$\text{AET} \approx 2\text{ET}_W - \text{RefET}$$

ET_w in this formulation is assumed to represent an AET rate for a comparable regional-scale wet surface with similar insolation and wind speed characteristics. Central to this relationship is that when the land surface is experiencing water stress, AET will decrease due to lack of available water, while RefET exhibits a similar complementary increase due to increasing sensible heat. This can lead to compound drought stress when

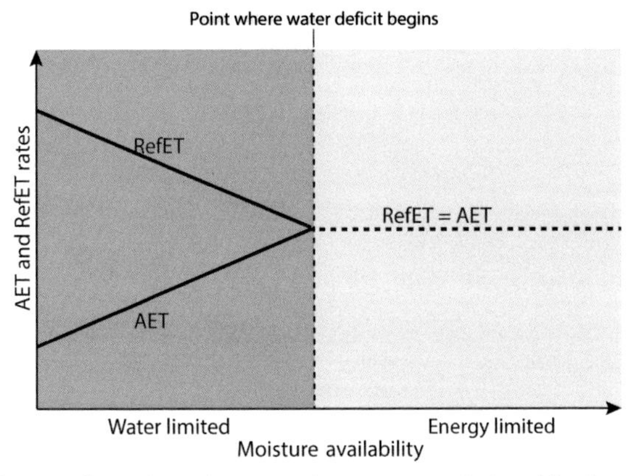

Figure 7.4 Schema depicting the complementary relationship between AET and RefET. *AET*, Actual evapotranspiration; *RefET*, reference evapotranspiration.

low water availability is accompanied by increased atmospheric water demands (Fig. 7.4).

7.7 Spatiotemporal variations in reference evapotranspiration and actual evapotranspiration, and their relationship to vulnerability and exposure

The complementary framework also provides a valuable conceptual tool for understanding the temporal and spatial evolution of drought, and how drought shock may interact with exposure and vulnerability to form a climate risk, as discussed in Chapter 6, Tools of the Trade 4—Mapping Exposure and Vulnerability. Most human and natural systems are adapted to the natural progression of rainy seasons. Setting aside regions that depend on wintertime precipitation (stored in snowpack and reservoirs), most areas depend heavily on the seasonal progression of summer rains, which tend to move north and south along with the passage of the Sun and the latitude of maximum insolation. Many such regions experience a seasonal progression (Fig. 7.5) from a water-limited state at the beginning of the season to an energy-limited state in the middle of the season, only to return to a water-limited state at the end of the rainy season. Before the rainy season, cloud-free skies bring copious solar radiation, and the lower atmosphere is typically quite dry, with $e_a << e_s$, leading to large VPDs. As the rains commence, clouds reduce the incoming radiation, while also bringing increases in moisture that reduce the VPD.

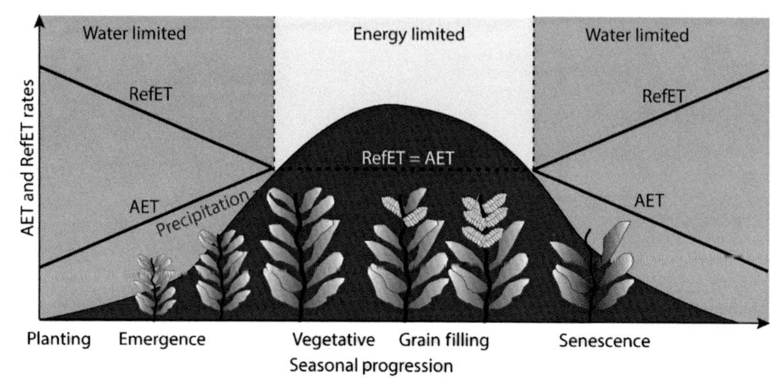

Figure 7.5 Schema depicting the complementary relationship between AET and RefET, along with a typical seasonal progression for a crop-growing region. *AET*, Actual evapotranspiration; *RefET*, reference evapotranspiration.

Assuming that the farmers' crop selection is well-aligned with the climate, it is typical to see crop emergence occur in the first water-limited stage of the growing season. These emergent plants require relatively little water. Then, as RefET decreases and precipitation and soil moisture increase, many annual plants dedicate energy from sugars and carbohydrates to putting on leaves. In this vegetative stage, plants focus on increasing their ability to absorb sunlight and transform it into more sugars and carbohydrates. Adding green biomass and the process of photosynthesis both require substantial amounts of water, and this requirement typically coincides with a period of time with, on average, high precipitation and relatively low RefET. Within this mid-season period, the overall hydrologic system is often energy-limited, because the moisture supply exceeds RefET.

Once cereal crops establish plentiful green biomass, they typically transition to a grain formation and filling stage (Fig. 7.5). The triggers for this transition, however, can vary substantially between crops and between different crop varietals. Some crops, such as maize, will transition based on accumulated "degree days." Other crops such as millet or sorghum may have triggers associated with the position of the Sun and the length of the local day.

When crops transition to the grain-filling stages, they dedicate substantial energy resources to the production of gene-carrying grains, that is, ears of corn, spikes of wheat, grains of rice, and pods of soy beans. After grain filling an annual grass or cereal plant's work is done. Energy from the Sun has been absorbed, combined with water and CO_2, and converted to seeds containing the next generation's potential growth and expansion. Leaves brown and drop off (senesce), and in this final stage of a crop's life cycle, anomalous wet conditions can also be harmful to farm production by making it harder to dry and harvest crops from the field.

The seasonal crop progression can be viewed both through a lens of exposure and vulnerability, as well as through a lens of relative water supply and demand. Here, we use the term "exposure" to refer to potential agricultural losses. Early in the season, vulnerability to rainfall deficits is fairly high, due to relatively high RefET conditions and relatively low average rainfall values. Exposure, however, may also be relatively modest. Plants that survive an early season drought may recover with bountiful mid-season rains. If the season is long enough, a failed planting may be replaced by another, more successful sowing. On the other hand, in areas where the growing season is short, a disruption at the onset of the season may make it very hard to recover, since the typical remaining period of adequate moisture supply (after the disrupted onset period) may be too short to support both the vegetative green biomass production and grain filling required to produce a successful harvest.

During the middle of the season, conversely, it is typical to find relatively low crop vulnerability, but relatively high crop exposure. This is especially true during the grain-filling stage. While infrequent in most good crop-growing areas, a transition to drought conditions, combined with feedbacks associated with the complementary relationship, can cause severe impacts just when crop exposure is highest. If soil moisture deficits lead to a reduction of AET, then the sensible heat flux must compensate to balance net radiation, which leads to an increase in air temperature. Since drought conditions are also typically accompanied by increased solar radiation—due to reductions in cloud cover—the increase in solar radiation also increases air temperature and RefET. These coupled processes often lead to increases in both RefET and air temperature. Increases in these terms can desiccate and damage crops, and such sensitivities can be especially critical during tasseling and grain-filling stages. Anomalous spikes in mid-season air or land surface temperatures can be a strong indication that rainfall deficits and increased atmospheric demand are likely to take their toll on crops.

These same general considerations can also be applied spatially. Consider Fig. 7.6, which sketches schematically (from left to right) a humid region with high precipitation and relatively low RefET, a semi-humid region with both moderate precipitation and RefET, and an arid

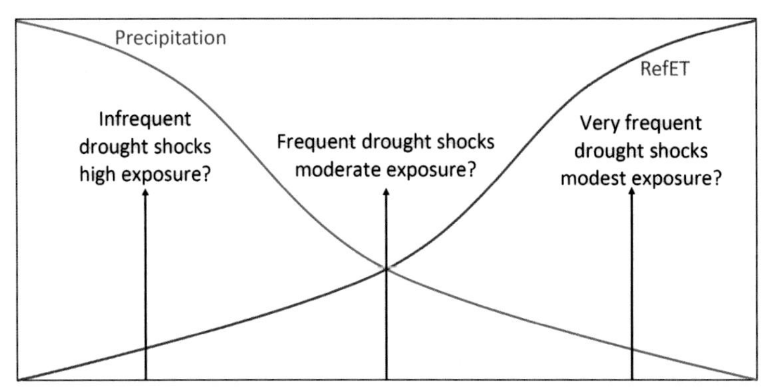

Figure 7.6 A conceptual representation of the interplay between spatial or temporal variations in precipitation and RefET. Because precipitation and RefET tend to be inversely correlated over space and time, there are often regions or periods with relatively high precipitation and low RefET, regions or periods with relatively low precipitation and high RefET, and regions or periods in the middle. Drought shocks in the first category might correspond to water deficits in highly product crop production regions, or water deficits during the critical vegetative and grain filling periods of crop growth. While less common, such shocks can be very damaging. Drought shocks in drier regions or time periods may be more frequent, but less damaging, since humans have typically adapted to these drier conditions. *RefET*, Reference evapotranspiration.

region with relatively low precipitation and relatively high RefET. In many cases, population densities and economic investments are greater for humid areas than for arid regions. On the left of Fig. 7.6, we might find the grain baskets of the world. On the right side of Fig. 7.6, we might find extremely food insecure pastoral or agropastoral communities. For the latter, two out of three seasons might be poor, and these fragile populations eke by on proceeds from that one infrequent good year. On the left, we might find a high degree of climatic stability. Most years are good, and agricultural vulnerability is relatively low. In between these extremes, we can identify a region with moderate climate vulnerability.

Effective DEWS need to be concerned with droughts in all three types of regions. Droughts in dryland areas can be devastating, and frequent, but the magnitude of overall exposure in these regions may be comparatively low, especially when overall population totals dependent on agricultural livelihoods in these arid lands are low. However, when characterized by high populations of pastoral and agropastoral communities, droughts in these regions can still lead to widespread food insecurity and even famine, which, in turn, require millions, or even billions, of dollars in emergency assistance. On the other hand, when droughts do strike humid areas with exposure, the associated losses can be very large. In 2012 drought struck almost the entire United States (cf. Fig. 2.5), while in 2015/16, a severe crop failure in the highly productive maize triangle in South Africa contributed to massive food insecurity across most of southern Africa. Understanding RefET and the role it plays in droughts can help us identify emerging water crises in all climate regimes.

References

Allen, R.G., Jensen, M.E., Wright, J.L., Burman, R.D., 1989. Operational estimates of reference evapotranspiration. Agron. J. 81, 650–662.

Bouchet, R.J., 1963. Evapotranspiration réelle et potentielle, signification climatique. IAHS Publ. 62, 134–142.

Hobbins, M., Wood, A., Streubel, D., Werner, K., 2012. What drives the variability of evaporative demand across the conterminous United States? J. Hydrometeorol. 13, 1195–1214.

Hobbins, M.T., Wood, A., McEvoy, D.J., Huntington, J.L., Morton, C., Anderson, M., et al., 2016. The evaporative demand drought index. Part I: linking drought evolution to variations in evaporative demand. J. Hydrometeorol. 17, 1745–1761.

Maes, W., Gentine, P., Verhoest, N., Gonzalez Miralles, D., 2019. Potential evaporation at eddy-covariance sites across the globe. Hydrol. Earth Syst. Sci. 23, 925–948.

Penman, H.L., 1948. Natural evaporation from open water, bare soil and grass. Proc. R. Soc. London. Ser. A: Math. Phys. Sci. 193, 120–145.

Thornthwaite, C.W., 1948. An approach toward a rational classification of climate. Geograph. Rev. 38, 55–94.

CHAPTER 8

Theory—indices for measuring drought severity

8.1 Introduction

This chapter will provide a general overview of some of the many possible indices used to monitor and quantify droughts. It should be noted at the outset that there are too many different potential drought indices to list them all here. Our purpose, rather, is to provide a general description of factors common to most drought indices and introduce a few of the most commonly used index categories. A more detailed discussion of drought indicators and drought indices[1] (Svoboda and Fuchs, 2016), which this chapter builds on, has recently been produced by the World Meteorological Organization in collaboration with the Global Water Partnership, the National Drought Mitigation Center, and the Integrated Drought Management Programme. Readers interested in a more detailed treatment may wish to consult this excellent resource.

Svoboda and Fuchs (2016) begin their description by highlighting the difference between drought *indicators* and drought *indices*. Indicators such as precipitation, temperature, runoff, soil moisture, snowpack, reservoir levels, or streamflow describe potential drought conditions. Indices are indicators that involve additional computations that can provide qualitative assessments of the severity, location, timing, and duration of events. In the broadest terms, most index values add qualitative information by either (1) reexpressing hydrologic information in the context of some type of water availability context, and/or (2) reexpressing drought information in terms of its historical context.

The latter transformation is conceptually more straightforward, so we will discuss it first. Droughts are typically infrequent events, and the dangers posed

[1] World Meteorological Organization (WMO) and Global Water Partnership (GWP), 2016: Handbook of Drought Indicators and Indices (M. Svoboda and B.A. Fuchs). Integrated Drought Management Programme (IDMP), Integrated Drought Management Tools and Guidelines Series 2. Geneva. <https://www.drought.gov/drought/sites/drought.gov.drought/files/GWP_Handbook_of_Drought_Indicators_and_Indices_2016.pdf>.

Drought Early Warning and Forecasting
DOI: https://doi.org/10.1016/B978-0-12-814011-6.00008-7

by droughts often increase rapidly as their *in*frequency increases. A 1-in-10 or 30- or 50-year drought will be much more intense, destructive, and dangerous, and carefully constructed high-quality monitoring systems can help rapidly identify such extreme events. Most human and environmental systems tend to be resilient to a frequent 1-in-3-year dry season. Few systems will be prepared to shrug off a 1-in-a-100-year drought. Assume that we have identified a drought within a given region and time period. In general, there will now be four typical approaches that can be used to express the "frequency" information associated with this specific drought. We can refer to the drought's rank, percentile value, return period, or standardized value.

Rank is one of the easiest expressions to understand. Ranks are derived from (1) taking all the indicator values associated with a region and time period, (2) sorting these values, and (3) determining the position (or rank) of the particular indicator of interest. Assuming the data is accurate, and the period of record adequate, such analysis can yield powerful statements such as "an analysis of rainfall data suggests that the recent drought is the worst on record" ... or "... second worst on record"... or "... third worst ...." These numbers (one, two, and three) represent the rank of the values from lowest to highest. Of course, the length of the time series plays an important role in how we might interpret this information. An exciting new satellite data set might only have 6 years of data. In this context, "lowest" might just mean below normal. On the other hand, in some contexts and countries, we might have more than a hundred years of information. In such a setting a rank of one, two, or three is very likely to indicate exceptionally dry conditions. This context is often provided by statements such as "During time period XYZ, region PDQ experienced the lowest rainfall/runoff/soil moisture/streamflow/snowpack/reservoir levels/vegetation health in MNO years."

While powerful and precise, statements such as these are hard to map and analyze operationally, especially when dealing with multiple regions, time periods, and indicator variables. One very common way to transform ranks is to reexpress them as percentiles. A percentile is simply the observed rank divided by MNO, where MNO represents the number of observations in the time series. For seasonal averages, this will be the number of years of data. Ranging from 1/MNO to 1, these percentile values describe the position of a particular event within the overall distribution of observed outcomes. Assuming MNO is measured in years, an event with a percentile of 1/MNO is a one-in-MNO years low event. A percentile of 0.5 corresponds to a once-every-other-year event.

A percentile value of 0.5 represents the median of our distribution, so we expect half the outcomes to be higher and half the outcomes to be lower. A percentile of one indicates a one-in-MNO-year high event. In general, caution should be applied when interpreting extreme values. By definition the tails of such distributions are sparsely populated. Nevertheless, the early and accurate identification of infrequent droughts, as opposed to frequent droughts, is a major objective and societal benefit provided by effective drought early warning systems (DEWS).

In addition to percentiles, another very common approach to expressing drought indices is to formulate them as standardized "*z*-scores." This approach builds on the fact that many people are familiar with the standard normal distribution, that is, the Gaussian normal distribution (bell curve) with an expected mean of 0 and an expected standard deviation of 1. The value of such a framework is an ease of interpretation. Values near 0 will be "typical." Values beyond ± 0.7 standardized anomalies will be abnormal. Values beyond ± 1.2 will be quite extreme. Values beyond ± 2 standardized anomalies will be exceptional.

There are two basic approaches used to calculate standardized anomalies—nonparametric and parametric. A nonparametric approach builds on the empirical percentile calculation described earlier. Data are ranked and expressed in percentiles by dividing by the total number of observations. These percentile values can then be directly transformed into standardized *z*-scores by using software to translate a percentile into the corresponding standard normal *z*-score values. For example, a percentile of 0.16 will produce a standard anomaly of -1. A -1 *z*-score corresponds to a 1-in-6-year low value. A percentile of 0.5 will be associated with a *z*-score of 0. These values also correspond with the median of the distribution. A percentile of 0.98 will be associated with a *z*-score of 2.0.

A very commonly used alternative "parametric" approach involves (1) fitting a statistical distribution to the observed data, then (2) using this distribution to identify theoretical percentile values, and then (3) translating the resulting percentile values into *z*-scores from a standard normal distribution with a mean of 0 and a standard deviation of 1. This parametric approach can help overcome deficiencies in the observed data set. For example, the limited observed time series may have a gap, resulting in two quite different sorted values being assigned quite similar ranks and percentile values. Extreme values may also not be captured well, given the short period of record of most historical hydrologic observations. In the parametric approach the drought analyst fits parameters to the observed distribution and then uses those parameters to transform the observed values into standardized z-scores from a normal distribution with a mean of 0 and a standard deviation of 1.

The simplest example of such a transformation involves the transformation of an observed value from a nonstandard normal distribution to a "standard normal" distribution with a mean of 0 and a standard deviation of 1. Some types of data, such as air temperatures and vegetation indices, are often well described by normal distributions. As an illustrative example, assume region A, in season B, has an average Normalized Difference Value Index (NDVI) of 0.4. Is this above or below normal? Exceptionally so or just a little? To immediately convey such information, we can calculate the historical mean (μ) and standard deviation (σ) of NDVI in this region and season. Now, we can express the observed value as a z-score: $Z_{\mathrm{NDVI}} = (\mathrm{NDVI}_{\mathrm{obs}} - \mu)/\sigma$. If $\mu = 0.6$ and $\sigma = 0.1$, we are looking at exceptionally low NDVI values, $Z_{\mathrm{NDVI}} = (\mathrm{NDVI}_{\mathrm{obs}} - 0.6)/0.1 = -2$ standardized anomalies. If $\mu = 0.6$ and $\sigma = 0.4$, we are looking at unexceptionally low NDVI values, $Z_{\mathrm{NDVI}} = (\mathrm{NDVI}_{\mathrm{obs}} - 0.6)/0.5 = -0.5$ standardized anomalies.

This general approach can be extended to transform variables with different, non-Gaussian distributions. For example, precipitation distributions tend to have skewed distributions with "fat tails." It is typically common to have many observations with low or no rainfall, and a few observations with extreme rainfall values. These distributions are often fit with three parameter conditions or gamma distributions. One parameter describes the probability of a nonrain event. The other two parameters (the shape and scale) describe the skewed distribution of the nonzero precipitation events. While the details may vary, in all cases, a parameterized distribution can be used to translate an observed value into a percentile between 0 and 1. Then, this percentile value can be translated into a z-score based on the standard normal Gaussian distribution.

This process, applied to precipitation, is used to produce one of the most popular indices: the Standardized Precipitation Index (SPI) (McKee et al., 1993). This index is extremely popular because (1) precipitation variations tend to be the single most important driver of most droughts, (2) it is very easy to interpret, (3) it is fairly easy to calculate, and (4) precipitation data are usually readily available. The World Meteorological Organization has a manual dedicated to the SPI.[2] The SPI can be easily calculated at a

[2] http://www.droughtmanagement.info/literature/WMO_standardized_precipitation_index_user_guide_ en_2012.pdf

range of temporal scales, which typically range from 5 days to as many as 12 or 24 months. Soil moisture will tend to respond to short-term SPI conditions. Water storage, groundwater, and streamflow may reflect longer term precipitation anomalies. In general, 1- to 2-month SPI values can be used to examine meteorological drought. The 1- to 6-month SPI is often used to assess agricultural drought. The 3- to 24-month SPI may be used to examine hydrologic drought.

Fig. 8.1 provides examples of SPI calculations. The top panel shows a typical gamma distribution for a shorter time period and/or a drier location. Such locations tend to have values near 0. Occasionally, but infrequently, relatively extreme values will be observed. Such distributions, which are very common precipitation distributions, will have median values that are lower than the mean. Typical values, characterized by the

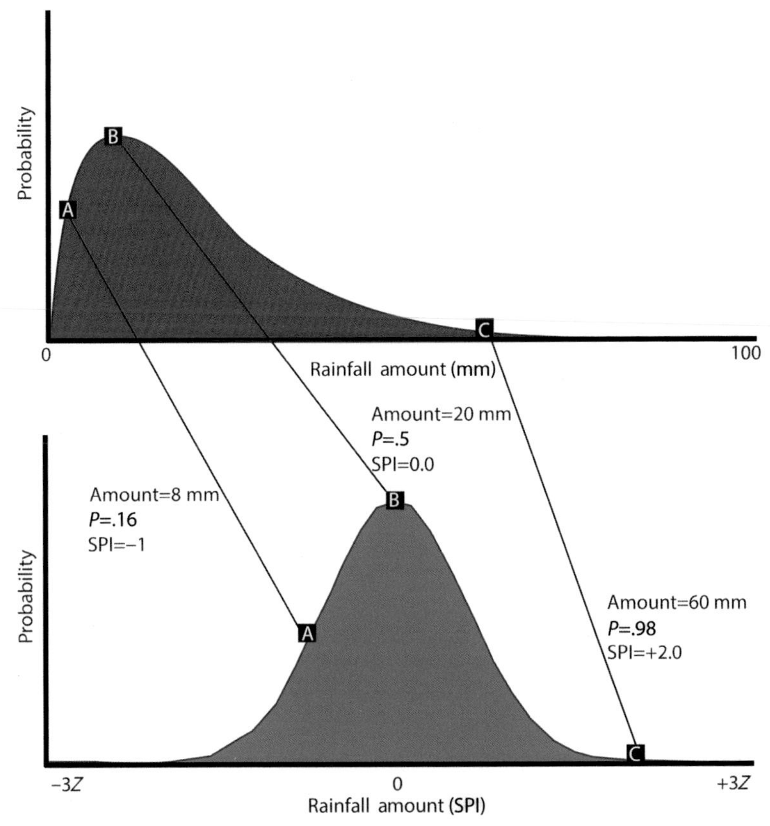

Figure 8.1 Examples of SPI calculations based on a gamma distribution (top) and standard normal distribution (bottom). *SPI*, Standardized Precipitation Index.

median, will be lower than the distribution's mean value, which will be pulled by a few wet events.

We can walk through the SPI calculation process for the example rainfall amounts: 8, 20, and 60 mm, marked A, B, and C in Fig. 8.1. The top panel of this figure displays a typical probability distribution function. The y-axis displays the probability of a certain rainfall amount. The x-axis shows the rainfall amounts ranging from 0 to 100 mm. These totals are characteristics of 10-day totals in semiarid growing regions. A value of 8 mm might be associated with a theoretical percentile value of 0.16. To transform this into a standard normal z-score the equations for a Gaussian standard normal distribution are used to calculate the corresponding value from the normal distribution (shown in the bottom panel). This would result in a value of −1 standardized anomalies. This process can be repeated for a value of 20 mm, corresponding to the median of the observed data set. The corresponding percentile value and associated standard normal z-score would be 0.5 and 0Z. The median of the Gaussian distribution is mapped to the mean of standard normal distribution. Our final value is a rainfall total of 60 mm, which we assume to be associated with a percentile value of 0.98. This would correspond with an SPI value of +2Z.

Advantages and disadvantages of "standardizing" index values: there are a number of different approaches to standardizing index variables using ranks, empirical percentiles, theoretical percentiles (based on parameterized distributions), and z-score-based indices such as the SPI. All of these approaches share certain advantages. First, they allow values in different climatic regions and different aggregation periods to be compared. Second, they express the observed conditions in terms of a probabilistic historic context. A historical database is used to determine how likely or frequent the observed values are, given previous observations. While these are substantial advantages, they also come with inherent disadvantages. Standardized anomalies can obscure important physical aspects of the systems being analyzed. For example, consider a −3 3-month SPI outcome for an extremely wet region. The rainfall variability in such a region might be very low compared to that region's mean rainfall. So, a −3 SPI value might still represent plenty of precipitation from an agricultural or regional water use perspective. Conversely, consider the case of a 3-month −3 SPI value from a region and time period that typically has almost no precipitation. Such SPI values may have limited physical meaning.

8.1.1 Consider multiple expressions of each individual data source

In addition to examining multiple sources of information, it is also highly recommended to examine each individual data source using multiple expressions. Standardized values such as SPI can be used to rapidly compare regions and seasons but also tend to generate a lot of potential false alarms associated with areas where precipitation variability is very low. This can draw the analyst's attention to out-of-season regions. One should also examine the individual data source in the original units and in terms of arithmetic (observation minus mean) or percent anomalies (100 × observation divided by the mean). When calculating percent anomalies, a small value is often included in the denominator to gracefully handle situations when the mean approaches 0. While the interpretation of absolute values and anomalies will be case-specific, this specificity can be powerful. For example, for an agricultural or pastoral outcome, an annual total precipitation of 200 mm is almost certain to be inadequate. Other indicator variables that have fairly direct physical interpretations [soil moisture, NDVI, or actual ET (AET)] also provide meaningful bases for drought detection.

8.2 Length of record and nonstationary systematic errors

Unfortunately, drought analysis can be very demanding when it comes to data set accuracy. We are often concerned about anomalies in hydrologic conditions in relatively dry areas; hence, relatively small errors can be problematic. This can be especially problematic when errors are systematic (i.e., nonrandom). While in practice it is hard to separate these terms, we can conceptually decompose a data source's error into two components: random error and systematic error. For this discussion, let us assume that random errors have a mean of zero and some nonzero variance, while systematic errors have a nonzero mean but zero variance. Consider a set of satellite-based rainfall estimates. They might systematically over- or underestimate rainfall at a given location by an amount expressed by the bias (β). They will also have a random error component that we can represent as ε_t, where t represents time. If we average over some N time steps, the expected value of the bias error will be $N\beta$. The systematic errors accumulate. This may not be a problem if the bias is relatively small compared to the mean and variance of the true precipitation distribution. Furthermore, expressing the satellite observations as either standard anomalies, arithmetic anomalies, or percent anomalies will remove the influence of this systematic error entirely. The random errors, even if quite large, will tend to cancel each other out over time. Some errors will be positive, some errors will be negative, and when these errors are added together, the variance of the composite error term will be substantially smaller. This is very good news for DEWS. Data sets do not need to be perfect. Averaging over longer time periods or regions will tend to increase our signal to noise ratios, making it easier to identify large anomalies. This cancellation, which is an expression of the central limit theorem in statistics, is good news for many data-sparse regions of the world.

On the other hand, nonstationary systematic errors can be huge problems for DEWS. If a data set's bias becomes nonstationary in an appreciable way, that is, β becomes a function of time, β_t, then the nonstationary systematic errors may become as large as the true year-to-year variations in the variable of interest. Averaging over space and time may accentuate these systematic errors, overwhelming the cancelation of local random error values.

Unfortunately, all three primary sources of drought indicator values, reanalyses, satellite observations, and interpolated station observations, may contain nonstationary systematic errors. Reanalysis systems often incorporate changing constellations of satellites and this can produce dramatic shifts in estimated precipitation, moisture, temperature, and reference evapotranspiration (RefET), especially in the latter part of the 1990s when microwave sounding units began providing estimates of atmospheric profiles. Shifts in satellite observing systems can also create inhomogeneities in satellite-based estimates of precipitation, vegetation, and soil moisture.

What is less well appreciated is that station-based data sets can also exhibit large nonstationary systematic errors. These errors arise because the networks of station observations change over time. As an example, consider Fig. 8.2. This figure schematically shows the outcome interpolating stations. In year 1 (top panel), we have an observation at a low elevation (near the ocean on the left) with a relatively low precipitation value (1). We also have a weather observation taken from a high elevation in the mountains on the right, with a relatively high value of 3. Assuming that we interpolate these values to a central equidistant location, we would expect that interpolated value to be 2 or (1 + 3)/2. The middle panel shows that in the next year, the same observations produce the same interpolated value in the central location. Then in year 3, the observation in the mountain stops recording or reporting. Our interpolation process now only finds one neighboring station, the "1" value near the coast. The interpolated precipitation estimate

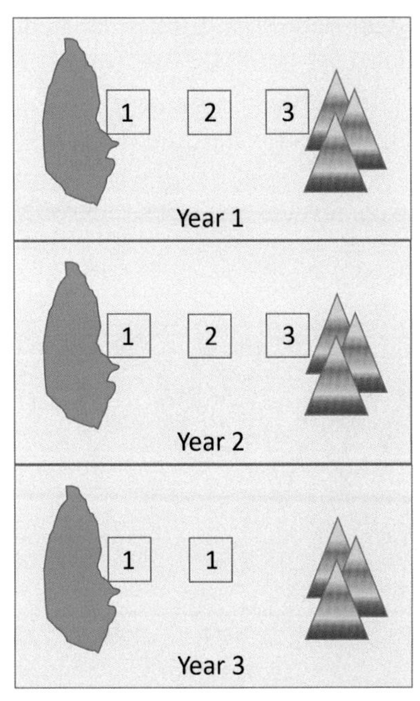

Figure 8.2 Schematic diagram explaining how interpolating precipitation values can produce nonstationary systematic errors when observation networks change over time.

in the central location is now "1." Has precipitation declined by 50%, producing a severe drought? Or have we have just witnessed the problematic nature of nonstationary systematic errors?

Nonstationary systematic errors can be reduced by improving intersatellite calibration efforts (Janowiak et al., 2001; Knapp et al., 2011; Yang et al., 2016), reanalysis assimilation processes (Robertson et al., 2011; Gelaro et al., 2017; Reichle et al., 2017), and by using station interpolation strategies that work with anomalies rather than "raw" station values (Becker et al., 2013; Schneider et al., 2017). Consider Fig. 8.2 again. If we assume that the means of the beachside and mountainside stations are 1 and 3, and that we interpolate anomalies and then add them to a background mean field, then the discontinuity shown in Fig. 8.2 would disappear. The expected value at the equidistant location would be the local mean, plus an interpolated anomaly of 0. Our phantom drought would disappear.

The limited availability of homogeneous, rapidly updated, and reasonably long period of record data sets led to the development of the Climate Hazards center InfraRed Precipitation with Stations (CHIRPS) data set (Funk et al., 2015a). This data set blends a high-resolution climatology

(Funk et al., 2015b) with stations and intercalibrated (Knapp et al., 2011) thermal infrared precipitation estimates. By construction, this data set has been designed to support drought assessments in data-sparse regions with complex terrain. Fig. 8.3 shows a validation study from Funk et al. (2015a).

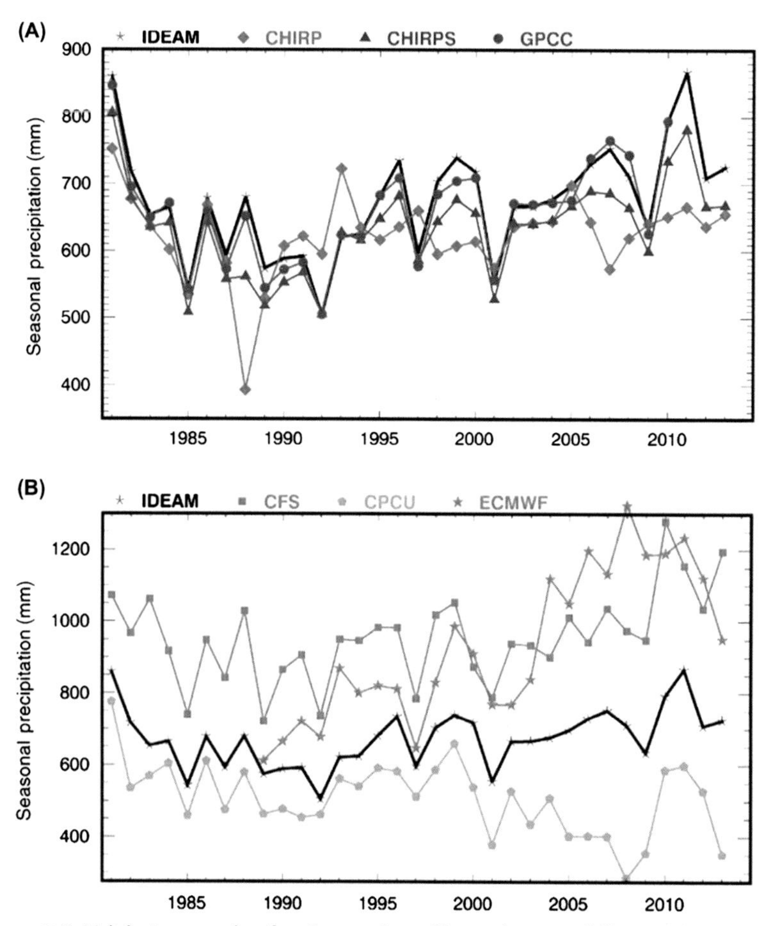

Figure 8.3 Validation results for September–November rainfall in Colombia. The black line in panels (A) and (B) shows average national rainfall based on a very dense set rain gauge observations provided by the Colombian meteorological agency. The brown values are averages based on a high-quality set of interpolated gauge observations produced by the Global Precipitation Climatology Centre (GPCC). The green and blue lines show precipitation from a satellite-only product (CHIRP) and a gauge-satellite blended product (CHIRPS). The time series in panel (B) show rainfall values from two reanalyses (the CFS and ECMWF) and an interpolated station data set produced by the Climate Prediction Center—the CPC Unified archive. *CHIRPS*, Climate Hazards center InfraRed Precipitation with Stations; *CFS*, Coupled Forecast System; *ECMWF*, European Center Medium range Weather and Forecasts.

Overall, the best performing data set was the interpolated gauge data produced by the Global Precipitation Climatology Centre (GPCC), shown in Fig. 8.3A. This is a gold standard data set based on the interpolated anomalies combined with a background precipitation climatology. The latency of this very high-quality data set, unfortunately, is too long to make it applicable to most early warning applications. While not perfect, the performance of the CHIRPS data set was fairly comparable to that of the GPCC. Note, however, that the performance of the satellite-only CHIRP was fairly poor in this part of the world. In Fig. 8.3B, we compare the Colombia Meteorological Agency data (IDEAM) with values from two reanalyses: the Coupled Forecast System (CFS) and the European Center Medium range Weather and Forecasts (ECMWF). While these time series tend to go up and down from year to year in good agreement with the IDEAM validation data set, they also exhibit substantial upward drifts that obscure droughts in the latter part of the time series.

Importantly, the station-only Climate Prediction Center Unified (CPCU)—interpolated gauge data set also exhibits a substantial downward drift after about 2002. Note also that the year-to-year performance also appears to degrade substantially after this point. Between 1981 and 2002 the highs and lows in the CPCU and IDEAM track fairly well. After 2002 they bear little resemblance. This CPCU data set is based on the interpolation of "raw" daily station observations. We suspect that a substantial decline in the number of gauges after 2002 led to a serious degradation of product performance.

Effective DEWS must consider the type of issues discussed here. Without good indicator data sets, droughts will likely be misidentified or missed entirely.

8.2.1 Frequently used satellite and "combination" drought indicators and indices

While the number of drought monitoring indicators and indices is vast, we provide here a partial list of some of the more commonly used products that go beyond precipitation and temperature. To keep this book from becoming dated too quickly, we describe the basics of several indicator data set categories but avoid details describing algorithms or satellites. Note also that essentially every indicator data set can be expressed as a z-score, rank, anomaly, or return period. Outputs from hydrologic models, reanalyses, or RefET calculations (discussed in previous chapters) can also be used as potential drought indicators.

Vegetation indices: Vegetation tends to absorb red light and emit infrared radiation, so the normalized difference in the satellite-observed radiation from these wavelengths can be used to define the Normalized Difference Vegetation Index, NDVI = (IR − Red/(IR + Red). The IR + Red term in the denominator tends to cancel out changes in the overall intensity of the incoming solar radiation. A slightly more complicated version (the Enhanced Vegetation Index) also uses emissions from the blue wavelengths to improve performance in heavily vegetated regions.

Microwave soil moisture estimates: Karthikeyan et al. (2017a,b) provides a good overview of current microwave soil moisture retrievals efforts. Satellite-based soil moisture retrievals are either based on passive (radiometer) or active (radar) microwave sensors. Active sensors send radar signals down to Earth's surface and then record the returning microwave radiation values. Passive sensors record radiometer brightness temperatures at microwave frequencies. Both types of retrieval systems rely on the fact that lower frequency microwave emissions from soils are influenced by the presence of soil moisture. Since microwave emissions are less energetic the spatial resolution of microwave imagery is generally coarser, typically on the scale of 25 km^2.

Estimates of actual evapotranspiration: There is a large and vibrant literature on the important topic of directly estimating AET from satellite data. The Surface Energy Balance Model (SEBAL) (Bastiaanssen et al., 1998) introduced the idea that the definition of hot and cold pixels with a remote sensing image could be used to estimate AET values. Assuming that the incoming radiation to both pixels is similar, this leads to the plausible conclusion that hot pixels are hot because they must offset the incoming radiation primarily through their own emission of thermal energy. Cold pixels, on the other hand, can be colder because they are also balancing the downwelling solar radiation via evaporation and transpiration. One widely used, globally available elaboration of the SEBAL concept is the Simplified Surface Energy Balance (SSEB) developed by Senay et al. (2011, 2013). The SSEB approach calculates a RefET value and then assumes that the coldest values will have AET values close to this amount. Conceptually, this approach is quite similar to the complementary relationship discussed in this chapter.

Another broadly similar approach is the Atmosphere-Land eXchange Inversion (ALEXI) model (Anderson et al., 1997; Anderson et al., 2011). The ALEXI model was developed as an extension of the two-source energy balance (TSEB) model of Norman et al. (1995). The TSEB assumes that the observed brightness temperature of a pixel is produced by a combination of a soil component and a vegetation component. This approach solves for the energy balance of each of these components, then solves for the total net radiation, ground heat flux, sensible heat flux, and AET terms. The ALEXI extension to the TSEB approach can be used in conjunction with ~5 km thermal infrared imagery to estimate AET at regional scales. This technique takes unique advantage of the rapidity of geostationary satellite observations of the land surface. Observations taken during the morning are used in conjunction with an atmospheric boundary layer model to estimate the magnitude of the sensible heat flux (Anderson et al., 1997). Net radiation, the ground heat flux, canopy evapotranspiration, and sensible heat flux values are then combined to produce an overall estimate of AET. An Evaporative Stress Index (ESI) can then be calculated by dividing the estimated AET by the estimated RefET. Values less than 1 indicate evaporative stress.

Standardized Precipitation–Evaporation Index (SPEI): The SPEI was developed by Vicente-Serrano et al. (2010) at the Instituto Pirenaico de Ecologia in Zaragoza, Spain. The SPEI is similar to the SPI but also uses estimates of RefET (often based on observed air temperatures). Parametric distributions fit to the differences between these two terms (precipitation minus RefET) allow the observed differences to be expressed as standardized *z*-scores. Like the SPI, the SPEI can be applied across many different temporal and spatial scales. The SPEI can account for the impact of temperature but may overemphasize the impact of temperatures on RefET, if a temperature-based RefET formulation is used. The SPEI, however, can be calculated using other more physically sound RefET formulations such as the Penman–Monteith algorithm, described in this chapter.

Crop water availability models: In general, there is a huge range in the complexity of currently available crop models. One model that is widely used in many developing nations is the Water Requirement Satisfaction Index (WRSI). Originally developed by the FAO (Doorenbos and Pruitt, 1977; Frère and Popov, 1979, 1986), the WRSI provides an indicator of crop performance by estimating the fraction of water availability over the course of a growing season. FAO research indicates that well-calibrated WRSI models typically exhibit linear relationships with crop yields. Spatially explicit (gridded) versions of the WRSI are now widely used (Verdin and Klaver, 2002).[3]

The calculation of the WRSI begins with a rainfall-based estimate of the start of the growing season. From that point forward to the end of the

[3] https://chc.ucsb.edu/tools/geowrsi

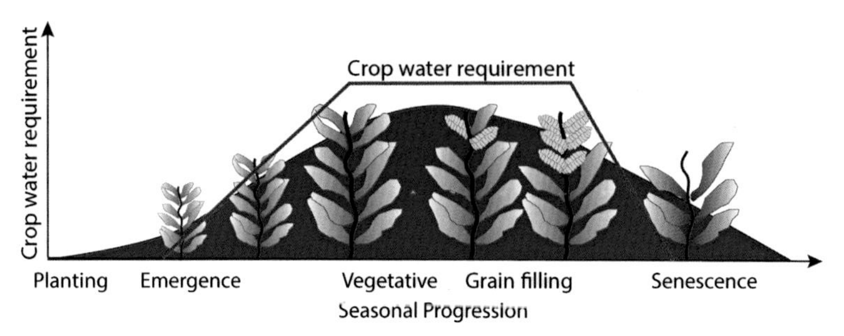

Figure 8.4 Seasonal progression of the WRSI crop water requirement. *WRSI*, Water Requirement Satisfaction Index.

growing season the model estimates a time series describing the plant water demand curve (Fig. 8.4). This time series describes a specific quantity of water (say 30 mm per 10-day time period) that the crop should extract from the soil column to maintain ideal growth. This quantity is quite low when the plants first emerge from the ground. The water requirements are high during the vegetative and grain-filling periods, but then decline rapidly. The water requirement estimates are based on time-varying crop coefficients multiplied by RefET. Increasing RefET increases the crops' water demands.

The offsetting term in the WRSI calculation is an estimate of the AET flux that the crops draw from the soil. The time-varying AET term is a function of the plant available for soil moisture and the crop water demand (WR). If the available soil moisture meets or exceeds the crop water demand, the AET will equal the crop water demand. The combination of AET, rainfall, and the water-holding capacity of the soil is used to provide a running estimate of the total available soil moisture. Running totals of crop AET and WR are calculated, and the WRSI at time t is provided by the following relationship:

$$\mathrm{WRSI}_t = 100 \times \frac{\sum_{i=1}^{t} \mathrm{AET}_t}{\sum_{i=1}^{t} \mathrm{WR}_t}$$

where $t = 1$ is assumed to be the beginning of the season. If AET always matches WR, the WRSI value will be 100. WRSI values of ~ 50 indicate severe crop water stress. Caution should be used when analyzing WRSI values early in the season. A good start to the season (with WRSI values near 100) may not be very indicative of late-season performance, because much of the crop water demand will arise later in the season.

The Palmer Drought Severity Index (Palmer, 1965; Alley, 1984) is another widely used index that combines precipitation and RefET with assumptions about soil water capacity to estimate plant water availability. At the complex end of the spectrum, there are many sophisticated crop models such as Decision Support System for Agrotechnology Transfer (DSSAT) and the Agricultural Production Systems sIMulator (APSIM). While these models typically require substantial calibration and detailed weather information, they can provide very accurate information about crop yields. Their complexity, however, tends to limit their utility as a general drought indicator.

8.3 Per capita water availability

The Falkenmark per capita water availability index, discussed in Chapter 6, Tools of the Trade 4—Mapping Exposure and Vulnerability, provides one example of an index that provides a quantitative (if rather simplistic) assessment of water supply compared to per capita water demand. This index can help translate an absolute quantity, say the total runoff from a river basin, into an index that may be more closely tied to drought impacts. Such an index will have an inherent spatial scale (a basin) but can also be examined at many different temporal scales. For example, we might be interested in the Falkenmark Index values during the months or weeks of the year with the lowest runoff. For some social and environmental systems, even a relatively brief loss of available water may be terribly disruptive. Such crises may also be accompanied by a rapid deterioration of water quality, potentially leading to outbreaks of diseases such as acute watery diarrhea or even cholera. On the other hand, long, persistent multiyear dryness can sap water storage reserves, creating serious problems even for nations with deep financial reserves and sophisticated water-management capacities.

8.4 Summary and discussion

As this chapter has illustrated, the topic of drought indicators and drought indices is extensive. Even if we just isolate a signal variable, such as precipitation, there are a myriad of ways to calculate and present a drought index. Practitioners should consider drought intensities both in the terms of standardized values, ranks, and/or percentiles, and in terms of the actual quantities being examined. At the end of the day, effective DEWS will need to target specific impacts. Such targeting can help select specific indicators and design effective triggers. The choice of spatial and temporal units of aggregation is important. For agricultural and rangeland monitoring purposes the timing of the selected analyses will be guided by the typical seasonal progression of precipitation and vegetative water demand. Agricultural and rangeland monitoring, however, do differ substantially, because crop outcomes are linked to their very specific phenology of grain-producing crops. Atypical late-season rainfall can often produce rapid improvements in rangeland conditions. For most rainfed crops, bountiful late-season rainfall is much less likely to overcome the negative impacts associated with a poor start to the season, or serious water deficits during the vegetative and grain-filling stages of crop growth. Hydrologic drought tends to respond to lower frequency water deficits and these responses can be very complex based on the specific region's water storage, transport, and use characteristics. Chapter 12, Practice—Actionable Information and Decision-Making Networks, will revisit this topic in an applied setting.

References

Alley, W.M., 1984. The Palmer Drought Severity Index: limitations and assumptions. J. Clim. Appl. Meteorol. 23 (7), 1100−1109.

Anderson, M.C., Norman, J.M., Diak, G.R., Kustas, W.P., Mecikalski, J.R., 1997. A two-source time-integrated model for estimating surface fluxes using thermal infrared remote sensing. Remote Sens. Environ. 60 (2), 195−216.

Anderson, M., Kustas, W., Norman, J., Hain, C., Mecikalski, J., Schultz, L., et al., 2011. Mapping daily evapotranspiration at field to continental scales using geostationary and polar orbiting satellite imagery. Hydrol. Earth Syst. Sci. 15 (1), 223−239.

Bastiaanssen, W.G.M., Menenti, M., Feddes, R.A., Holtslag, A.A.M., 1998. A remote sensing surface energy balance algorithm for land (SEBAL). 1. Formulation. J. Hydrol. 212−213, 198−212.

Becker, A., Finger, P., Meyer-Christoffer, A., Rudolf, B., Schamm, K., Schneider, U., et al., 2013. A description of the global land-surface precipitation data products of the Global Precipitation Climatology Centre with sample applications including centennial (trend) analysis from 1901-present. Earth Syst. Sci. Data 5, 71−99.

Doorenbos, J., Pruitt, W., 1977. Crop water requirements. FAO irrigation and drainage paper 24, Land and Water Development Division, 144. FAO, Rome.

Frère, M., Popov, G., 1979. Agrometeorological Crop Monitoring and Forecasting. FAO.

Frère, M., Popov, G., 1986. Early Agrometeorological crop yield forecasting. In: FAO, Plant Production and Protection Paper (73).

Funk, C., Peterson, P., Landsfeld, M., Pedreros, D., Verdin, J., Shukla, S., et al., 2015a. The climate hazards infrared precipitation with stations—a new environmental record for monitoring extremes. Sci. Data 2.

Funk, C., Verdin, A., Michaelsen, J., Peterson, P., Pedreros, D., Husak, G., 2015b. A global satellite assisted precipitation climatology. Earth Syst. Sci. Data Discuss. 7, 1–13.

Gelaro, R., McCarty, W., Suárez, M.J., Todling, R., Molod, A., Takacs, L., et al., 2017. The Modern-Era Retrospective Analysis for Research and Applications, Version 2 (MERRA-2). J. Clim. 30 (14), 5419–5454.

Janowiak, J.E., Joyce, R.J., Yarosh, Y., 2001. A real-time global half-hourly pixel-resolution infrared dataset and its applications. Bull. Am. Meteor. Soc. 82, 205–217.

Karthikeyan, L., Pan, M., Wanders, N., Kumar, D.N., Wood, E.F., 2017a. Four decades of microwave satellite soil moisture observations: Part 1. A review of retrieval algorithms. Adv. Water Resour. 109, 106–120.

Karthikeyan, L., Pan, M., Wanders, N., Kumar, D.N., Wood, E.F., 2017b. Four decades of microwave satellite soil moisture observations: Part 2. Product validation and inter-satellite comparisons. Adv. Water Resour. 109, 236–252.

Knapp, K.R., Ansari, S., Bain, C., Bourassa, L.M.A., Dickinson, M.J., Funk, C., et al., 2011. Globally gridded satellite (GriSat) observations for climate studies. Bull. Am. Meteorol. Soc. 92 (7), 893–907.

McKee, T.B., Doesken, N.J., Kleist, J., 1993. The relationship of drought frequency and duration to time scales. In: Proceedings of the Eighth Conference on Applied Climatology, Anaheim, CA, American Meteorological Society Boston, MA.

Norman, J.M., Kustas, W.P., Humes, K.S., 1995. Source approach for estimating soil and vegetation energy fluxes in observations of directional radiometric surface temperature. Agric. For. Meteorol. 77 (3), 263–293.

Palmer, W.C., 1965. Meteorological Drought, Research Paper No. 45. U.S. Weather Bureau, Washington, DC, p. 58.

Reichle, R.H., Liu, Q., Koster, R.D., Draper, C.S., Mahanama, S.P., Partyka, G.S., 2017. Land surface precipitation in MERRA-2. J. Clim. 30 (5), 1643–1664.

Robertson, F.R., Bosilovich, M.G., Chen, J., Miller, T.L., 2011. The effect of satellite observing system changes on MERRA water and energy fluxes. J. Clim. 24 (20), 5197–5217.

Schneider, U., Finger, P., Meyer-Christoffer, A., Rustemeier, E., Ziese, M., Becker, A., 2017. Evaluating the hydrological cycle over land using the newly-corrected precipitation climatology from the Global Precipitation Climatology Centre (GPCC). Atmosphere 8 (3), 52.

Senay, G.B., Budde, M.E., Verdin, J.P., 2011. Enhancing the Simplified Surface Energy Balance (SSEB) approach for estimating landscape ET: validation with the METRIC model. Agric. Water Manage. 98 (4), 606–618.

Senay, G.B., Bohms, S., Singh, R.K., Gowda, P.H., Velpuri, N.M., Alemu, H., et al., 2013. Operational evapotranspiration mapping using remote sensing and weather datasets: a new parameterization for the SSEB approach. JAWRA J. Am. Water Resour. Assoc. 49 (3), 577–591.

Svoboda, M., Fuchs, B., 2016. Handbook of drought indicators and indices. Integrated Drought Management Tools and Guidelines Series 2. I. D. M. P. (IDMP). World Meteorological Organization (WMO) and Global Water Partnership (GWP), Geneva.

Verdin, J., Klaver, R., 2002. Grid-cell-based crop water accounting for the famine early warning system. Hydrol. Process. 16 (8), 1617–1630.

Vicente-Serrano, S.M., Beguería, S., López-Moreno, J.I., 2010. A Multiscalar Drought Index sensitive to global warming: the Standardized Precipitation Evapotranspiration Index. J. Clim. 23 (7), 1696–1718.

Yang, W., John, V., Zhao, X., Lu, H., Knapp, K., 2016. Satellite climate data records: development, applications, and societal benefits. Remote Sens. 8 (4), 331.

CHAPTER 9

Sources of drought early warning skill, staged prediction systems, and an example for Somalia

In 1948 the mathematician, philosopher, and founder of information theory Norbert Wiener (1894–1964) wrote a brilliant collection of essays, *Cybernetics: or Control and Communication in the Animal and the Machine.* Wiener originated the concept of cybernetics, a field of information theory focused on the development of intelligent behavior arising through feedbacks. Whether arising across communications with newfangled computers such as the electronic numerical integrator and computer (ENIAC) or across burgeoning global telephone transmission lines, transmission errors threatened the very core of the mid-20th century computation and communication. Defeating transmission errors such as droughts can be very difficult. Typically, most data are good, and we just want to find the bad bits. Just as typically, most weather is normal, and we want to identify the droughts that arise in a few isolated locations and times. To address this problem of error detection, Wiener introduces ideas that relate to the modern cyber-security concept of defense-in-depth (DiD), in which a series of defensive mechanisms are layered in order to protect valuable data or information (Fig. 9.1, left). DiD is a multilayered approach with *intentional* redundancies. Just like a pitchfork-bearing prole outside a castle, attackers are faced with consecutive lines of defense—the city walls, then the moat, then the ramparts, and then finally, the central keep.

Applying the DiD approach to drought early warning systems (DEWS) (Fig. 9.1, right) can result in very effective and timely early warning. Multiple distinct data sources provide multiple opportunities to catch potential disasters. Multiple opportunities are provided to communicate risk. A nuanced staged approach builds on the strengths of long-range forecasts but does not rely too much on these less certain sources of information. A DiD approach to DEWS development takes advantage of the same logic that signal processor pioneers such as Wiener applied to error detection—redundancy in a DEWS is a good thing. Assume that

Drought Early Warning and Forecasting
DOI: https://doi.org/10.1016/B978-0-12-814011-6.00009-9

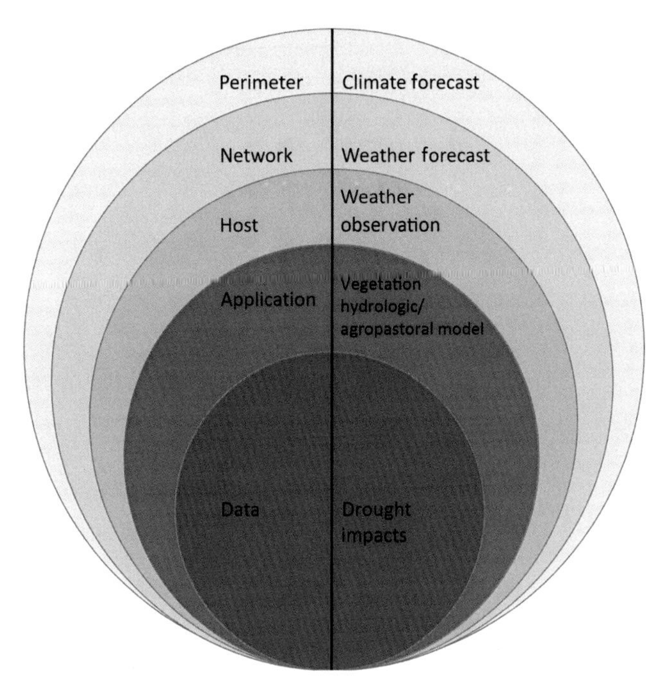

Figure 9.1 A defense-in-depth approach to designing drought early warning systems. Sequential sources of climate, weather, and land surface information are used in a way that provides both redundancy and optimal increasingly impact-specific information.

you have a filter *A* that can detect and remove 50% of the error in a signal. Then, let us say that filter *B* can also remove 50% of the error. Does applying *A* and then *B* remove 75% of the error? If *A* and *B* detect and remove the same errors, no. *A* would remove 50% of the errors and *B* would detect nothing. But if *A* and *B* are independent detection systems, *A* would detect half of the original errors, and *B* would detect half of the remaining errors—reducing the overall error rate to just 25% of the original. If additional independent 50% filtering filters *C* and *D* are then applied, the fraction of the original errors would drop to 12.3% and 6.2%.

Such logic underlies the phased sequential convergence of evidence (PSCOE) approach displayed in Fig. 9.1. Multiple independent sources of information are used, providing multiple opportunities to spot crises. One clear advantage of a PSCOE approach is that it provides support for a staged system of alerts that takes advantage of the unique capabilities of multiple sources of information. Drought early warning skill arises from the slow variation of ocean temperatures, predictable variations in weather, and persistent anomalies in vegetation and soil moisture (Fig. 9.2).

9.1 The ocean as a source of skill

Let us begin with the ocean. The global oceans vary at a much slower tempo than the atmosphere, because the ocean is much denser than air. It takes far more energy to heat or move a cubic meter of water than a cubic meter of atmosphere. At the surface of the ocean a cubic meter of water weighs about 1017 kg. At 15°C and sea level the density of air is 1.2 kg. The ocean is 1000 times denser. The specific heat capacity of the ocean [3850 J $(kg\ °C)^{-1}$] and atmosphere [1158 J $(kg\ °C)^{-1}$] describes how much energy would be required to heat these fluids by 1°C. The specific heat capacity of water (H_2O) is caused by hydrogen bonds between water molecules. It takes about $1017 \times 3850 \approx 3.9$ million Joules to heat 1 m^3 of sea water by 1°C. It takes $1.2 \times 1158 \approx 1.4$ thousand Joules to heat the same volume of air by 1°C. It takes about 2000 times the amount of energy to heat water as it does air. The oceans, therefore, possess a large "thermal inertia." Since it takes a lot of energy to change ocean temperatures, ocean temperature anomalies tend to persist from week to week and month to month. This persistence, combined with our ability to model some important types of ocean–atmosphere interactions, such as those related to the El Niño–Southern Oscillation (ENSO), provides the foundation for statistical and dynamic climate forecasts based on observed ocean conditions. Dynamic climate forecast models are discussed in Chapter 4, Tools of the Trade 1—Weather and Climate Forecasts.

Climate forecasts typically advance on monthly timescales and provide the greatest forecast lead times. These forecasts, however, also come with the highest levels of uncertainty and provide information at coarse spatial scales. Coupled ocean–atmosphere models may also have trouble representing local weather mechanisms. They tend to perform better in some regions and not so well in others. Multimodel ensembles can also sometimes exhibit spurious certainty. For example, the ocean–atmosphere feedbacks leading to the development of El Niño events are very difficult to model, and current climate models tend to overpredict the certainty of El Niño formation. If 90% of the model simulations predict a strong El Niño, or rainfall deficits in Southern Uganda in June, 6 months in the future, this may overrepresent the true certainty of these assessments. Climate processes such as the ENSO system typically involve coupled

Figure 9.2 Effective DEWS utilize multiple sources of predictive skill. Skill arises from the slow variation of ocean temperatures, predictable variations in weather, and persistent anomalies in vegetation and soil moisture. *DEWS*, Drought early warning systems.

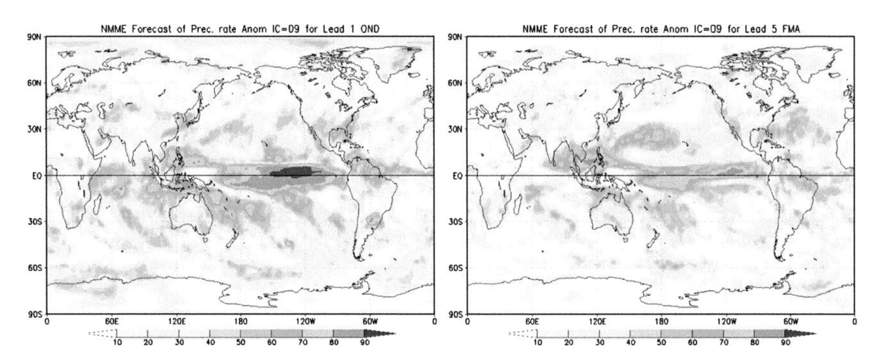

Figure 9.3 NMME forecast skill maps, based on September forecasts for October–November–December (left) and February–March–April (right). One hundred percent indicates perfect forecast skill. *NMME*, North American Multi-Model Ensemble. *Obtained from https://www.cpc.ncep.noaa.gov/products/NMME/ on September 8, 2019.*

ocean–atmosphere feedbacks that can be highly nonlinear and difficult to model (Ferrett and Collins, 2016). This can lead to a tendency for the frequency of El Niño to be overpredicted. Forecast certainties should be based on comparisons with historical data, and such assessments can typically be accessed on sites hosting multimodel forecast ensembles, such as the site maintained for the North American Multi-Model Ensemble (NMME).[1] In general, skills for precipitation and temperature forecasts will be much higher at shorter lead times and over the tropical Pacific Ocean, where ENSO dominates seasonal variability (Fig. 9.3).

In addition to observed and predicted sea surface temperature (SST) conditions, experienced climate analysts pay close attention to how the tropical atmospheric circulation is responding to a given set of SST anomalies. Atmospheric responses depend on the overall disposition of ocean temperatures, not just the local SST anomalies in a fixed region, such as the Niño3.4 box (5°S – 5°N, 170°E – 150°W) often used to depict El Niños. On the other hand, when strong El Niño and La Niña-like SST gradients are accompanied by vigorous anomalies in Indo-Pacific winds, atmospheric water vapor, and precipitation, an extreme ENSO state should provide opportunities for prediction. In a warming world, such extreme states induce exceptionally warm SST anomalies, which can be related to severe droughts in teleconnected regions. For example, the 2015–16 El Niño and the following 2016–17 La Niña produced a series

[1] https://www.cpc.ncep.noaa.gov/products/NMME/

of droughts that helped propel more than 50 million sub-Saharan Africans into severe food insecurity (Funk et al., 2018). First, a climate change-enhanced El Niño contributed to severe droughts in Ethiopia and southern Africa (Funk et al., 2016, 2017). Then human-induced warming of the Western Pacific contributed to severe back-to-back east African droughts in October–December of 2016 and March-to-May of 2017 (Funk et al., 2019a,b). While very damaging, from a climate hazards' perspective, these extreme SST states and associated impacts can be viewed as opportunities for prediction (Funk et al., 2019a,b).

9.2 Skill from persistent atmospheric conditions

The next source of predictive skill comes from persistent atmospheric conditions. While there are intrinsic chaotic aspects to atmospheric circulations, there is also considerable predictability on 1- to 14-day timescales, and efforts are being made to extend such weather forecasts to seasonal-to-subseasonal predictions on the 15- to 30-day timeframe. While limited in lead time, weather forecasts often have quite high levels of skill. Fig. 9.4 shows one such assessment for Africa. This figure shows the correlation between 10-day precipitation forecasts from Global Ensemble Forecast System[2] (GEFS) and 10-day observations of rainfall from the CHIRPS2.1 data set.[3] The GEFS forecasts are based on the ensemble average. While a detailed discussion of these results is beyond the scope of this chapter, we can note that rainfall correlations tend to be higher in places that are in season. Correlations are also a little higher, in general, in April than in October.

When combined with near real-time precipitation observations, weather forecasts can be very powerful tools for rapidly identifying mid-season droughts. Precipitation deficits tend to be leading indicators of soil moisture deficits, crop stress, and vegetation deficits. Combining observations and weather predictions near the middle of a growing season can provide timely advance warning. Furthermore, weather and climate forecasts can be used as inputs to hydrologic and crop models, providing a means of assimilating multiple sources of information to assess likely hydrological, agricultural, and pastoral impacts. The topic of integrating

[2] https://www.ncdc.noaa.gov/data-access/model-data/model-datasets/global-ensemble-forecast-system-gefs

[3] https://www.chc.ucsb.edu/data/chirps

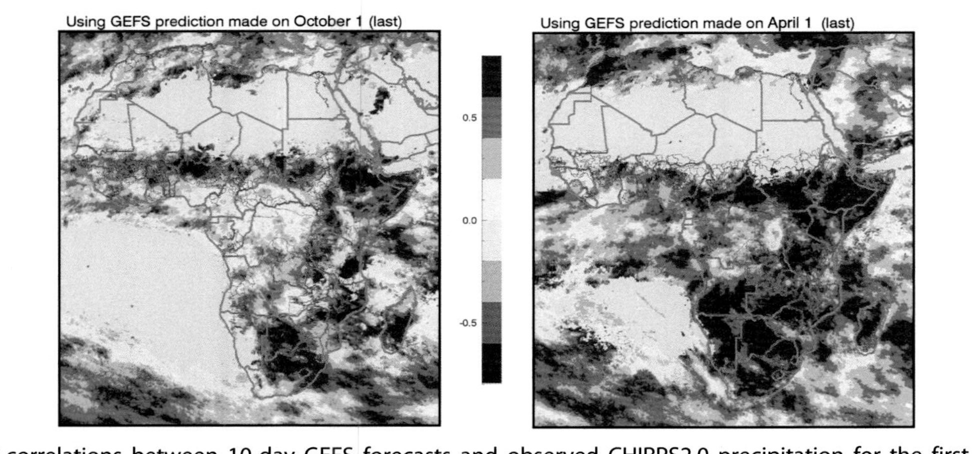

Figure 9.4 Maps of correlations between 10-day GEFS forecasts and observed CHIRPS2.0 precipitation for the first 10 days of October (left) and April (right). *CHIRPS*, Climate Hazards center InfraRed Precipitation with Stations; *GEFS*, Global Ensemble Forecast System.

observations and forecasts is explored in more depth in Chapter 11, Practice—Integrating Observations and Climate Forecasts.

9.3 Predictive skill from the land surface

The land surface provides a third source of predictive skill. While some forecast skill can arise through complex interactions between climate and land surface conditions, we focus here on the more straightforward application of lagged relationships between moisture supply, land surface response, and the associated drought impacts. The first precedes the second, which precedes the third, creating opportunities for effective prediction of impacts. The utilization of such lagged relationships can provide some of the most effective tools in a drought analyst's toolkit. In general, water balance deficits, arising through below-normal precipitation and above-normal evapotranspiration, result in lagged (~1 week to 1 month) deficits in soil moisture. Deficits in soil moisture can decrease evapotranspiration, causing land surface and 2 m air temperatures to increase, because the energy associated with the decreased evapotranspiration must be balanced by increases in the land surface emission temperatures and sensible heat fluxes. Dry soils and hot conditions during the first two-thirds of a growing season will inhibit plants' translation of CO_2 into carbohydrates and sugars by means of photosynthesis. For grain crops, this usually occurs first in a vegetative growth phase in which crops put on green leafy biomass (further facilitating photosynthesis) and a later "grain-filling" phase in which energy (sugars and carbohydrates) are channeled into the production of germ-bearing grains, that is, the part of corn, wheat, rice, and sorghum that we eat. Vegetation stress and deficits in plant biomass production tend to lag soil moisture deficits and temperature extremes, which, in turn, tend to lag water supply indicators (reduced precipitation and enhanced reference evapotranspiration values). Deficits in soil moisture and crop conditions can be simulated with hydrologic and crop models. Vegetation stress and deficit in plant biomass can be observed from space using vegetation indices. All three of these indicators (hydrologic simulations, crop simulations, and satellite-observed vegetation conditions) will tend to lead to agricultural outcomes, realized as grain harvests several months after the peak of the season. Thus all the sources of information described here provide effective predictive skill vis-à-vis agricultural crop production.

We can help contextualize this progression by discussing a specific country—Zimbabwe (Fig. 9.5). Like most regions in southern Africa, Zimbabwe experiences a single rainy season associated with monsoonal

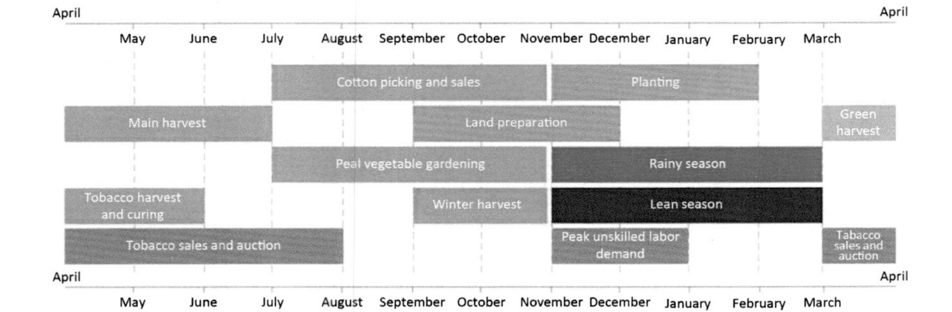

Figure 9.5 Seasonal progression of Zimbabwe food production and consumption. *Courtesy: From http://fews.net/southern-africa/zimbabwe on September 9, 2019.*

rains in austral summer. Rains tend to begin in earnest in November, peak in January, and diminish by March. Harvests typically occur between April and July. Ironically, hunger tends to peak around the time that the rains begin, as last year's food stores are diminished. The lean rainy seasons are contemporaneous (Fig. 9.5). Poor harvests extend the lean period, creating dangerous disruptions to lives and livelihoods. Evaluations of Zimbabwe crop statistics, precipitation, soil moisture values, and vegetation indices suggest, roughly, the following sensitivities. This sequence is common to many rainfed agricultural regions. Rainfall acts as a leading drought indicator, and Zimbabwe crop production is most closely related to rainfall in December and January (Funk and Budde, 2009), when many maize farms enter the grain-filling stage. Farmers tend to plant so as to harmonize crop water needs with the typical peak periods of precipitation. February soil moisture values appear to be strongly related to crop production, and finally, March–April vegetation index values are strong indicators of agricultural crop production. Since all three of these information sources (precipitation, soil moisture, and vegetation indices) lead the April–July harvest, all three sources can provide skillful forecasts. As illustrated later, combinations of mid-season precipitation observations, downscaled precipitation forecasts, and observations of Normalized Difference Vegetation Indices (NDVIs) can be combined in powerful ways to provide very effective mid-season alerts. In general, as one moves deeper into a given season, both the level of uncertainty and spatial specificity of the drought warning indicators increase. Staged and nuanced response systems can take advantage of this increasing accuracy and specificity by developing and implementing increasingly specific and targeted responses.

It should be noted, however, that the exposure and vulnerability of a given household, community, nation, or regional food economy may vary substantially from season to season. These prior conditions may amplify or diminish the impact of a specific seasonal drought. Examples can range from global to local. For example, the global food price spikes of 2008–11[4] increased the fragility of millions of households all over the world (Brown et al., 2015). Conflict in places such as Syria, South Sudan, and Yemen can greatly exacerbate the influence of poor agricultural outcomes. While such influences can travel through myriad pathways to erode resilience, there is one common prior influence that DEWS are well situated to recognize: the negative influence of past droughts. Such

[4] http://www.fao.org/worldfoodsituation/foodpricesindex/en/

influences may be most easily recognized in bimodal rainfall areas. Receiving rains twice a year, such areas are prone to damaging back-to-back droughts. Unimodal precipitation regimes can also be negatively impacted by poor outcomes in prior years. Consecutive droughts can drain aquifers, reservoirs, water holes, household savings, and government assets. When analyzing potential outcomes, drought analysts should consider the potential impacts of prior hydrologic and agroclimatic shocks.

9.4 Staged opportunities for prediction support defense-in-depth

A recurring theme throughout this chapter and this book is that the most effective drought early warning typically involves staged alerts and actionable decision support services. Such staged systems take advantage of multiple sources of information, providing multiple opportunities to catch a crisis. Furthermore, staged alerts provide response agencies and disaster risk reduction specialists with adequate time to put contingency plans in place (Choularton, 2007) and provide adequate long-term food security outlooks (Magadzire et al., 2017). To return to our DiD castle metaphor, we can imagine a royal counselor providing a series of pronouncements announcing that a rabble of pitchfork-waving invaders has formed beyond the postern gate ... passed across the moat ... breached the walls ... and finally entered the central keep. These successive and increasingly precise alerts provide governments and humanitarian agencies adequate time to respond.

Here, with reference to Somalia and a sequence of alerts provided in 2016 and 2017 by the authors, we sketch potential staged opportunities for prediction (Fig. 9.6). These staged opportunities begin with potential long-lead climate forecasts (1), which are only available for regions with strong teleconnections in time periods when strong climate forcing is likely or underway. The next window occurs as the rainy season commences (2). At this stage the thermal inertia and persistence of the ocean create a fairly high degree of certainty surrounding the like mid-season climate conditions. By mid-season (3) the drought analyst has a rapidly increasing arsenal of information. Precipitation conditions in the first-third to second-third of the season provide very valuable insights into the likely seasonal outcomes. Weather forecast models tend to perform well and offer additional insights into likely mid-season moisture supplies. Soil moisture and vegetation have also begun to respond to the water

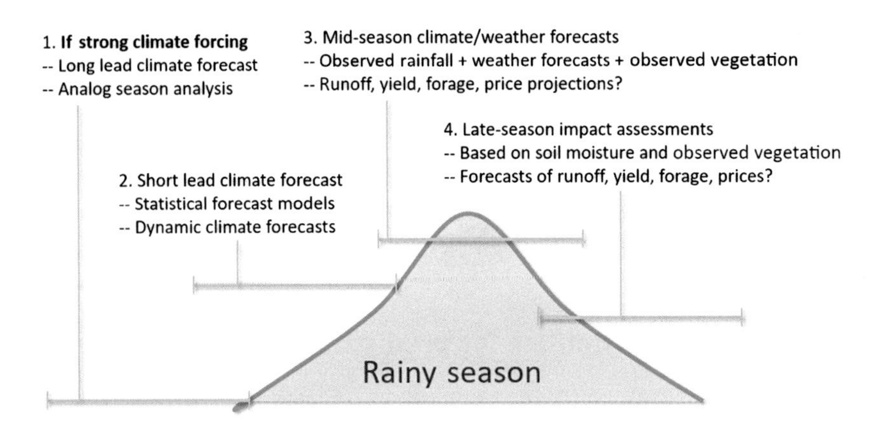

Figure 9.6 Staged opportunities for prediction pool sources of predictive skill and offer defense-in-depth. Note that practice historical data would be required at each stage to develop effective warning indicators.

availability and offer additional metrics of seasonal performance. Finally, at the close of the season, soil moisture, runoff, and vegetation provide detailed impact-relevant evidence of aridity, as well as indicators of yield, prices, food availability, and food access.

9.4.1 Stage 1: long-lead climate forecast

By definition, most of the time, oceanic SST conditions are characterized as normal. Occasionally, however, climate extremes arise, and these extremes are typically associated with climate modes such as the ENSO (Ropelewski and Halpert, 1987) and the Indian Ocean Dipole (Saji et al., 1999; Saji and Yamagata, 2003). La Niña ENSO phases, associated with eastern equatorial Pacific Ocean conditions, are often associated with dry conditions over Eastern Africa. Negative Indian Ocean dipole conditions, which are associated with cool SSTs near coastal East Africa and warmer than normal ocean conditions in the eastern equatorial Indian Ocean, are also associated with dry conditions over eastern East Africa.

La Niña events often follow moderate-to-strong El Niño events, creating an opportunity for El Niño-related droughts in places such as Ethiopia and southern Africa, followed by sequential La Niña-related back-to-back droughts in East Africa in October—December and then March—May (Funk et al., 2018). These repetitive back-to-back droughts can be devastating to food insecure farmers and pastoralists in Somalia.

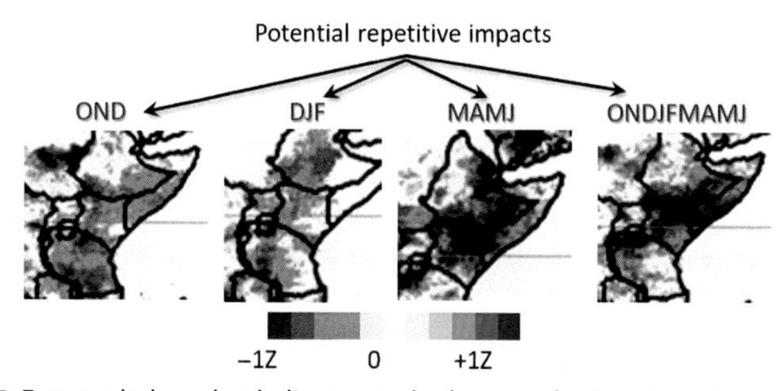

Figure 9.7 Extremely long-lead climate outlook example. *Courtesy: Based on analysis presented to USAID on May 19, 2016 showing standardized precipitation anomalies,* if *a moderate-to-strong La Nina formed.*

In May 2016, El Niño conditions had rapidly diminished, and according to the National Oceanic and Atmospheric Administration (NOAA) Climate Prediction Center, there was a 75% chance that a moderate-to-strong La Niña would develop by the fall of 2016. Prior Famine Early Warning Systems Network (FEWS NET), research indicated that *if* such an event occurred, climate change-related warming in the Western Pacific was likely to enhance the drought impacts of this La Niña. The FEWS NET science team held a colloquium in Washington, DC in mid-May, presenting composites of recent moderate-to-strong La Niñas (Fig. 9.7). What these slides emphasized was that *if* a moderate-to-strong La Niña was to develop, such conditions were likely to produce dry conditions during the following October–December, December–February, and March–June rainy seasons—constituting a major threat to eastern East Africa, in general, and Somalia, in particular, because the country is so arid, food insecure and dependent on pastoral livelihoods.

It should be noted that opportunities for long-lead forecasts are infrequent and typically related to ENSO-related SST extremes. Drought analysts should be on the lookout for such forcing scenarios. La Niñas, interacting with human-induced warming in the Western Pacific, may induce droughts across East Africa and significant portions of the northern hemisphere (Hoell et al., 2013, 2014). Moderate-to-strong El Niño drought impacts can affect Central America, Amazonia, Ethiopia, southern Africa, India, Southeast Asia, as well as northern Australia, and the Maritime Continent.[5]

[5] These types of composites can be created using the GeoCLIM tool https://www.chc.ucsb.edu/tools/geoclim/.

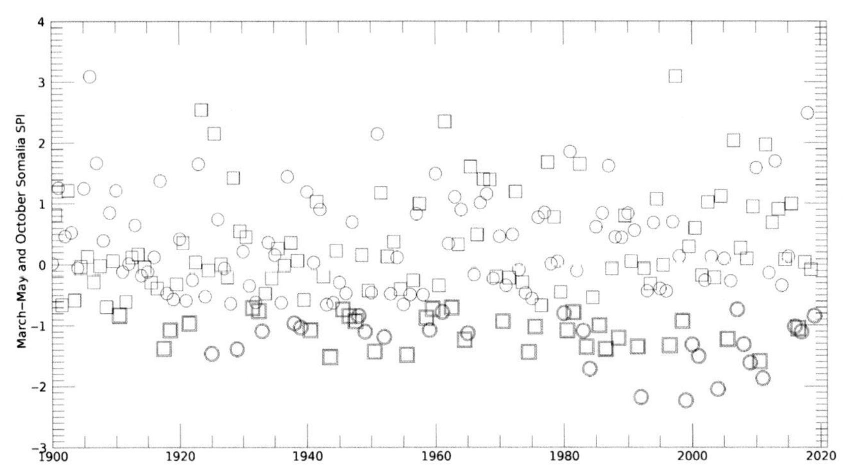

Figure 9.8 March–May (*circles*) and October–December (*squares*) Somalia SPI values. Values less than −0.7 SPI drawn in red. *SPI*, Standardized Precipitation Index.

9.4.2 Stage 2: short-lead climate forecasts

Short-lead climate forecasts are plausible for many regions of the world, due to the persistence on SSTs on 1- to 3-month timescales. Ocean conditions observed today are likely to be similar to conditions in a month or two. Dynamic climate forecasts, discussed in Chapter 4, Tools of the Trade 1—Weather and Climate Forecasts, provide an extremely useful tool for rapidly identifying likely precipitation outcomes everywhere in the globe. Care should be taken, however, to carefully assess the skill of these forecasts. Statistical forecasts, tailored to a given region and season, can sometimes outperform dynamic predictions. These statistical forecasts, however, should also be guided by an understanding of the relevant climate dynamics that is often informed by climate simulations. Another advantage of statistical forecasts is that they can be fairly easily implemented and are transparent in their assumptions. However, a downside of statistical forecasts is that they, unlike dynamic forecasts, have difficulty interpreting novel climate conditions.

Here, we produce an example very similar to analysis provided for an outlook on October 19, 2016 by the Climate Hazards Center (CHC).[6] As context, consider Fig. 9.8, which shows a long time series of Somalia March–May and October–December precipitation, expressed as Standardized Precipitation Index (SPI) values (McKee et al., 1993; Husak

[6] http://blog.chc.ucsb.edu/?p = 10

et al., 2007). These long time series are derived by combining 1900–80 Centennial Trends gridded station data (Funk et al., 2015a) with CHIRPS2.0 blended satellite-gauge data (Funk et al., 2015b). By construction, these data sets are built around the same climatology (Funk et al., 2015c), and well correlated for their period of overlap ($r = 0.87$ MAM, 0.97 OND). A plot of the MAM and OND SPI values is shown in Fig. 9.8. SPI values of less than −0.7 are plotted with red outlines.

Along with eastern Ethiopia and Central-Eastern Kenya the Somalia March–May rains have declined substantially, likely due to human-induced warming in the Western Pacific Ocean (Funk et al., 2018, 2019a, b). Since 1999, 11 out of 21 March–May rainy seasons have been below normal (SPI > −0.7). October–December rains do not exhibit a decline, probably because of warming in the Western Indian Ocean (Liebmann et al., 2014) and different upper level atmospheric responses over the Indo-Pacific warm pool (Funk et al., 2018). The interaction of climate change, manifested as exceptionally warm SSTs in the Indo-Pacific and naturally occurring La Niña conditions, can trigger frequent multiseason drought events. These events have also tended to follow strong El Niño events (Funk et al., 2018). The 1997/1998 El Niño transitioned into a strong La Niña in late 1998. The 1998 October–December, 1999 March–May, 2000 March–May, and 2001 March–May seasons were very dry. In 2006 a moderate El Niño was followed by La Niña conditions and dry March–May seasons in 2008 and 2009. In 2009–10 a moderate-to-strong El Niño transitioned into strong La Niña conditions, and the October–December 2010 and March–May rains were exceptionally dry. In 2016 a strong El Niño transitioned into a moderate La Niña, accompanied by exceptionally warm west Pacific SSTs. The October–December 2016 and March–May 2017 seasons were also very dry, compounding impacts from an unexpectedly dry 2016 March–May season.

While many factors impact East African rains, research has suggested that the persistence of west Pacific SSTs and La Niña-like SST gradients over the October–December and March–May rainy season can cause back-to-back droughts (Hoell and Funk, 2013). Designing early warning systems that can effectively capture these risks may help manage these repetitive climate shocks. It should be recognized, however, that there will be "surprise" droughts, such as the poor 2016 and 2019 March–May rainy seasons, that will probably not be predicted by most current statistical or dynamic models. A DiD early warning approach, therefore, provides multiple opportunities to identify emerging drought crises, greatly increasing the overall chance of success in projecting drought severity and impact throughout the season.

To develop a short-lead Somalia October–December rainfall forecast,[7] we begin examining the correlation between 1996 and 2018 September NOAA Extended Reanalysis SST (Huang et al., 2017) and Climate Hazards center InfraRed Precipitation with Stations (CHIRPS) October–December Somalia SPI. Fig. 9.9 shows a map with these correlations. A relatively short time period was chosen because recent research has indicated substantial nonstationarities in East Africa teleconnections (Nicholson, 2015, 2017). What we see in Fig. 9.9 is reassuring; we find a very strong negative teleconnection between SSTs in the Indo-Pacific warm pool and moderately negative correlations between the Western Indian Ocean and eastern equatorial Pacific. These patterns are consistent with prior research emphasizing ENSO and the Indian Ocean Dipole as key drivers of interannual variability. SSTs averaged over the warm pool region shown in Fig. 9.9 (90°E – 160°E, 15°S – 10°N) are very strongly anticorrelated with Somalia rains ($r = -0.79$). Eastern Pacific (180° W – 120°W, 5°S – 5°N) and Western Indian Ocean (45°E – 55°E, 5° S – 5°N) have positive correlations of 0.60 and 0.56. The warm pool exhibits anticorrelations with the east Pacific SSTs ($R = -0.66$) and Western Indian Ocean ($R = -0.41$). To develop a predictive model, these three SST time series were standardized, and used in a multivariate regression. The three slope coefficients from this model were −0.78, 0.00, and 0.32, indicating that the Warm Pool and Western Indian Ocean alone were the best predictors. The Warm Pool is the region of very warm waters surrounding Indonesia. This does not imply that the east Pacific SST (and implicitly ENSO) does not matter, but rather that the Warm Pool time series alone appears to effectively capture this information. The Western Indian Ocean–Warm Pool SST gradient will be associated with modulations in the Walker Circulation and low-level winds over the Indian Ocean (Hastenrath et al., 2010).

Take-one-away cross-validation was then used to assess out-of-sample forecast accuracy. This process involved holding out each year's data, fitting the model with the rest of the data, and then using the resulting regression coefficients to estimate the excluded year's Somalia rainfall value. The standard deviation of the year-to-year variations in the slope

[7] This analysis could be replicated using the GeoCOF tool https://www.chc.ucsb.edu/tools/geocof/.

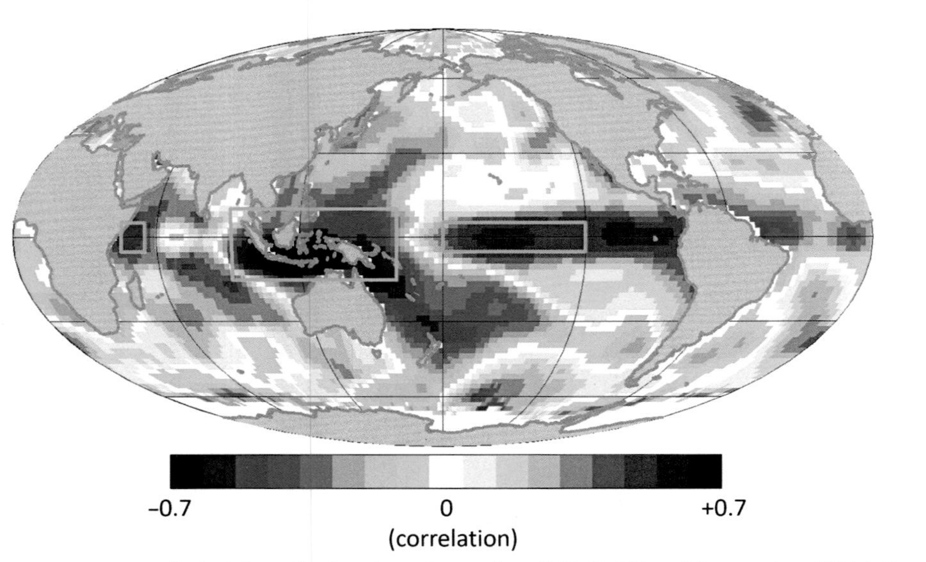

Figure 9.9 Correlation between 1996 and 2018 Somalia October–December CHIRPS SPI and September NOAA Extended Reconstruction SST. Green rectangles delineate regions used in statistical forecast models. *CHIRPS*, Climate Hazards center InfraRed Precipitation with Stations; *NOAA*, National Oceanic and Atmospheric Administration; *SPI*, Standardized Precipitation Index; *SST*, sea surface temperature.

coefficients was very small (~0.03Z/Z), indicating robust results. The cross-validated correlation remains high (0.78), with an overall standard error of 0.73Z. The model captures four of the five driest rainy seasons fairly well (1996, 1998, 2010, 2016) but misses the dry 2005 season (Fig. 9.10, left). The 2005 season was driven by an extreme Indian Ocean Dipole event, and not well captured by the model used here. The model, over the training period, did not produce any false alarms. While this does not preclude such an outcome in the future, the results presented here indicate that a simple robust model can perform quite well.

We next briefly evaluate the skill of dynamic coupled ocean–atmosphere climate forecasts based on September forecasts of October–December Somalia rains. The forecasts used here are drawn from the NMME. The simulations were downloaded from the International Research Institute. The model performance was very strong, with a correlation of 0.8 (Fig. 9.10, right). The models also did a good job of predicting the 2005 event, resulting in good detection performance for all four of the driest recent events.

9.4.3 Stage 3: mid-season climate/weather forecasts

We next provide an example of mid-season assessments that combine CHIRPS2.0 satellite-gauge rainfall estimates, CHIRPS-compatible

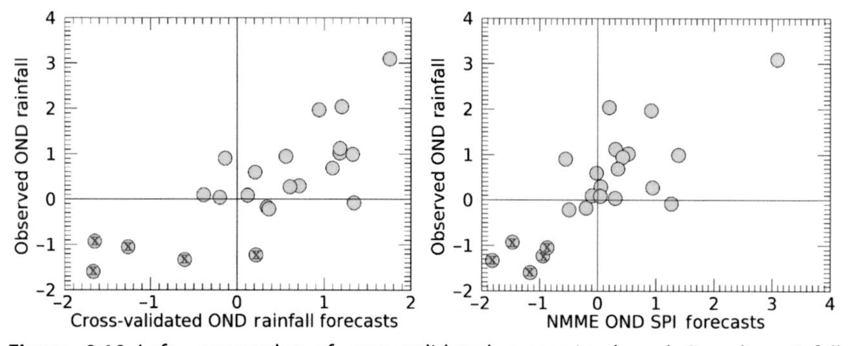

Figure 9.10 Left—scatterplot of cross-validated regression-based Somalia rainfall prediction based on warm pool and Western Indian Ocean SSTs. Right—scatterplot of climate model predictions and observed Somalia rainfall. Prediction based on ensemble average from the NMME. The cross-validated R^2 value of the statistical model (left panel) was 0.63. "X"s have been added to identify the five driest seasons in the period evaluated. *NMME*, North American Multi-Model Ensemble; *SST*, sea surface temperature.

downscaled GEFS precipitation forecasts, and observed vegetation conditions, represented here by eMODIS NDVI images. Two scenarios are evaluated. The first is based on October CHIRPS and NDVI data, along with the first dekad of November CHIRPS–GEFS precipitation forecasts. The two rainfall totals (observed and forecast) are combined (totaled). The second scenario is similar but uses October and November observations of CHIRPS and NDVI, and CHIRPS–GEFS forecasts for the first 10 days of December. Both sets of data are compared with October–December NDVI. This example is carried out at a national scale, but subnational results' applications could be developed in a straightforward manner.

The scatterplot in the left panel of Fig. 9.11 shows cross-validated regression results based on (1) October CHIRPS + dekad one of November CHIRPS–GEFS, (2) October NDVI, and (3) both predictor time series. Despite the early date of the analysis, high cross-validated correlation skills are identified (0.77, 0.70, 0.84, respectively). The two poorest seasons (2010 and 2015) are identified surprisingly well. The next two poorest seasons (2017 and 2018) are also identified well. It is quite concerning that 2016, 2017, and 2018 had such low October–December NDVI values. This behavior, however, seems to have been predicted reasonably well by the model.

Advancing 1 month, and carrying out a similar analysis (Fig. 9.11, right), we see an increase in forecast accuracy. The rainfall, NDVI, and combined rainfall and NDVI models had cross-validated correlation values of 0.78, 0.94, and 0.96. The strong performance of the just-NDVI model suggests that at a national scale, the NDVI signal has largely become fixed by early December. The relative lack of improvement in the forecast skill associated with the just-rainfall forecast may indicate that other factors are also helping to drive NDVI variations, that is, variations in air temperatures, reference ET, vapor pressure deficits, et cetera. While there is a substantial improvement in performance between the October and October-plus-November models, much of this improvement occurs in the upper right quadrant of this scatterplot. Behavior in November helps quantify the differences between mediocre and above-normal seasons.

Overall, these results emphasize the high degree of predictability associated with both short-lag climate forecasts (Fig. 9.10) and midseason monitoring (Fig. 9.11). By early December, poor national-level outcomes can be assessed with a high degree of certainty, and even 1 month into the rainy season we find surprisingly high detectability of

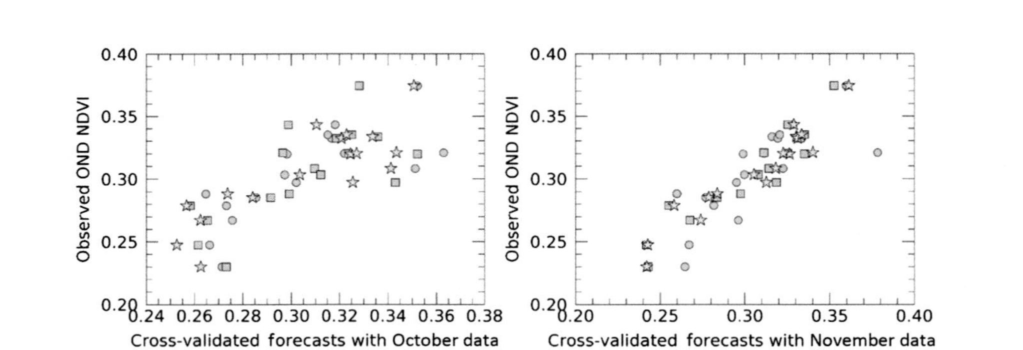

Figure 9.11 Left—scatterplot of October–December Somalia NDVI based on observed October rainfall and NDVI and predicted rainfall for the first dekad of November. Cyan circles depict forecasts of observed NDVI based on precipitation. Pink boxes represent forecasts of future NDVI based on early-season NDVI observations. Finally, yellow stars present forecasts based on both precipitation and NDVI. Right—same but for October–November observations and the first dekad of December CHIRPS–GEFS weather forecasts. *CHIRPS*, Climate Hazards center InfraRed Precipitation with Stations; *GEFS*, Global Ensemble Forecast System; *NDVI*, Normalized Difference Vegetation Index.

poor "short" rainy season outcomes (Fig. 9.11, left). Both statistical and dynamic models appear to capture the associated rainfall deficits quite well (Fig. 9.10).

9.4.4 Stage 4: late-season impact assessments

At the close of a growing season the early warning analyst has the most available information. Even after the rains have stopped, however, there is still a great deal of opportunity for prediction, since the impact of poor harvests, pasture conditions, and terrestrial water supplies are likely to be felt for months to come. At this stage the analysis of high-resolution satellite imagery can be very powerful. Using this data to monitor large regions can be prohibitive, but once problem regions have been identified, high-resolution data can be used to confirm and refine impact assessments. At this stage, of course, field assessments are also a critical source of information. We do not carry out a detailed example here, but rather refer to a 2016 analysis carried out by scientists at the European Commission's Joint Research Center (JRC) in support of an international joint alert[8] underscoring potential famine conditions in Somalia. This statement reflected a shared view of current conditions and the likely evolution of the situation in Somalia by major actors involved in global food security monitoring and early warning: The European Commission's JRC, the FEWS NET, the Food and Agriculture Organization of the United Nations, and the United Nations World Food Programme. Food security assessments indicated that more than 2.9 million people were facing crisis and emergency food security conditions until June of 2017. Some 363,000 children under 5 were acutely malnourished.

To underscore the very poor crop production prospects for the October–December, collaborators from the JRC used high-resolution satellite imagery to provide detailed maps of fields, which have either not been planted or where crops have failed in the most productive areas in the Lower Shabelle river basin (Reproduced as Fig. 9.12). As can be seen in Fig. 9.12, in 2016 only irrigated fields in the proximity of the Lower Shabelle river show green vegetation (*green areas*) at the end of the deyr season, while more peripheral irrigated areas and rainfed areas to the north and to the south are completely dry (*orange areas*). In early 2017, this analysis helped reinforce January harvest estimates indicating extremely low levels of crop production in Southern and Central Somalia (75% below 5-year average).

[8] https://documents.wfp.org/stellent/groups/public/documents/ena/wfp290554.pdf

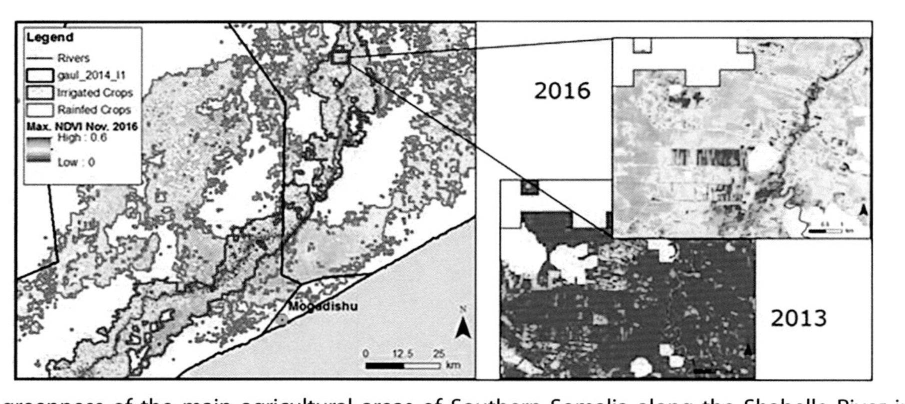

Figure 9.12 Vegetation greenness of the main agricultural areas of Southern Somalia along the Shabelle River in November 2016. The orange areas are dry fields that are normally at the peak of the crop cycle in this season. The green belt (left) in the middle corresponds to irrigated fields next to the Shabelle River. Less than half of the irrigated area (*purple polygon*) is actively producing in 2016. To the right the expanded views of an irrigated area show that, except for some larger fields next to the river, most plots are dry. The same area is also shown for a normal season (2013). *Courtesy: JRC, with Landsat eight data processed in Google Earth Engine, land cover provided by the FAO's Somalia Water and Land Information Management (SWALIM) project.*

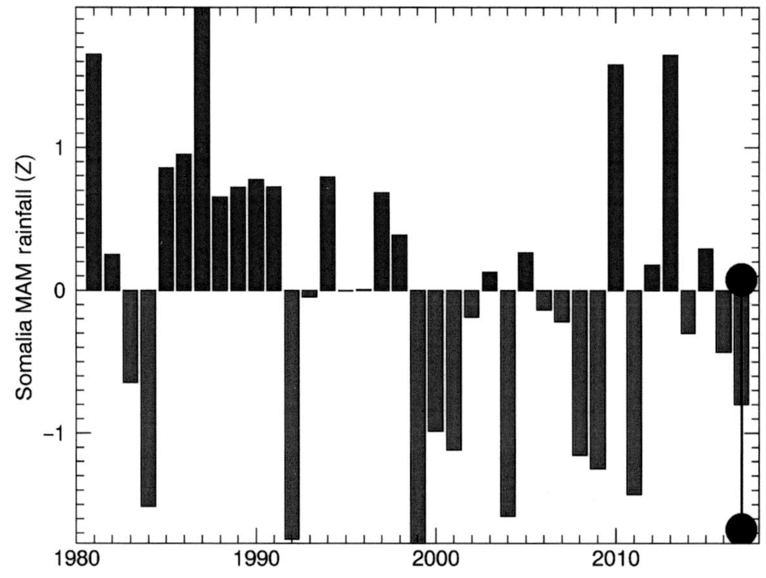

Figure 9.13 Time series of observed 1981–2016 Gu rains, together with a statistical forecast for 2017. The statistical forecast was based on a cross-validated regression model. The black line and circles denote the 2017 forecast and 80% confidence intervals. *Courtesy: WFP (2017) Persistent drought in Somalia leads to major food security crisis, Multi-Agency Joint Alert, https://documents.wfp.org/stellent/groups/public/documents/ena/wfp290554.pdf.*

9.5 Summary: staged strategies for effective early warning

We have now worked through examples of all four of the staged opportunities for prediction shown schematically in Fig. 9.6. When the La Niña was first predicted in late spring of 2016, long-lead outlooks were used to alert decision-makers to potential problems (Fig. 9.7). After the La Niña had commenced and just before the October–December season began, September ocean conditions provide skillful sources of prediction (Fig. 9.11). Advancing just 1 month, we find that rainfall and satellite observations provide a surprising level of skill that increases even more the following month (Fig. 9.11). Finally, after the season, high-resolution satellite data provide a detailed basis for examining agricultural conditions (Fig. 9.12). We briefly conclude with yet one more long-lead assessment, based on an assessment by the CHC[9] carried out in support of the Joint

[9] http://blog.chc.ucsb.edu/?p = 148

Alert (see footnote 7). This statistical forecast for the 2017 March–May rains, based on January sea-surface temperatures, is presented in Fig. 9.13. This forecast was based on research explicitly carried out to support East African food security decision-making in situations similar to those of 2010/11. Since the 1980s, March–May rains in Somalia have declined precipitously, with poor rainfall levels in most years since 1999. While good rains were received during the El Niño-like 2010 and 2013 March–May seasons, January 2017 sea-surface conditions (Fig. 9.4) were similar to those associated with most recent March–May dry seasons—warm west Pacific sea-surface temperatures appear alongside cool or neutral conditions across the equatorial Eastern Pacific. The associated statistical forecast (Fig. 9.3) was for a $-0.8Z$ standardized anomaly $\pm 1.1Z$. Taken together, this sequence of staged alerts helped motivate early and effective humanitarian assistance in Somalia in 2017.

9.6 Conclusion

Droughts are slow onset disasters. Their gradual nature can make them challenging to detect and predict, but this slow evolution also provides multiple opportunities for drought identification and mitigation. The most sophisticated approaches to drought early warning employ different types of products at different drought stages. Before the season, statistical and dynamic climate models can help anticipate outcomes, as can the state of various large-scale climate modes such as the ENSO. As the season progresses, 1- to 2-week weather model predictions can provide useful information, especially if these forecasts can be interoperable with observations. At the middle and end of a typical growing season, soil moisture estimates from land surface models or vegetation health indices such as the NDVI provide high-quality assessments of conditions. While these products lag precipitation, they provide more meaningful information about actual surface conditions. Hence, they tend to be very useful in the latter half of a rainy season. Understanding the relative utility of these products, when and where each can be used most effectively, and how they can be used together, provides a solid foundation for effective drought monitoring and prediction.

References

Brown, M., et al., 2015. Climate Change, Global Food Security, and the U.S. Food System. 146 pp.

Choularton, R., 2007. Contingency Planning and Humanitarian Action: A Review of Practice. Humanitarian Practice Network.

Ferrett, S., Collins, M., 2016. ENSO feedbacks and their relationships with the mean state in a flux adjusted ensemble. Clim. Dyn. 1–20.

Funk, C., Budde, M.E., 2009. Phenologically-tuned MODIS NDVI-based production anomaly estimates for Zimbabwe. Remote Sens. Environ. 113 (1), 115–125.

Funk, C., Nicholson, S.E., Landsfeld, M., Klotter, D., Peterson, P., Harrison, L., 2015a. The Centennial Trends Greater Horn of Africa precipitation dataset. Sci. Data 2 (150050).

Funk, C., Peterson, P., Landsfeld, M., Pedreros, D., Verdin, J., Shukla, S., et al., 2015b. The climate hazards infrared precipitation with stations—a new environmental record for monitoring extremes. Sci. Data 2.

Funk, C., Verdin, A., Michaelsen, J., Peterson, P., Pedreros, D., Husak, G., 2015c. A global satellite assisted precipitation climatology. Earth Syst. Sci. Data Discuss. 7, 1–13.

Funk, C., et al., 2016. Assessing the contributions of local and east Pacific warming to the 2015 droughts in Ethiopia and Southern Africa. Bull. Am. Meteorol. Soc. 97, S75–S80.

Funk, C., et al., 2017. Anthropogenic enhancement of moderate-to-strong El Niños likely contributed to drought and poor harvests in Southern Africa during 2016. Bull. Am. Meteorol. Soc. 37.

Funk, C., Harrison, L., Shukla, S., Pomposi, C., Galu, G., Korecha, D., et al., 2018. Examining the role of unusually warm Indo-Pacific sea surface temperatures in recent African droughts. Q. J. R. Meteorol. Soc. 144, 360–383.

Funk, C., Hoell, A., Nicholson, S., Korecha, D., Galu, G., Artan, G., et al., 2019a. Examining the potential contributions of extreme 'Western V' sea surface temperatures to the 2017 March-June East African Drought. Bull. Am. Meteorol. Soc. 100 (1), S55–S60.

Funk, C., et al., 2019b. Recognizing the Famine Early Warning Systems Network (FEWS NET): over 30 years of drought early warning science advances and partnerships promoting global food security. Bull. Am. Meteorol. Soc. Available from: https://journals.ametsoc.org/doi/full/10.1175/BAMS-D-17-0233.1?mobileUi = 0.

Hastenrath, S., Polzin, D., Mutai, C., 2010. Circulation mechanisms of Kenya rainfall anomalies. J. Clim. 24 (2), 404–412.

Hoell, A., Funk, C., 2013. Indo-Pacific sea surface temperature influences on failed consecutive rainy seasons over eastern Africa. Clim. Dyn. 43 (5–6), 1645–1660.

Hoell, A., Funk, C., Barlow, M., 2013. The regional forcing of Northern hemisphere drought during recent warm tropical west Pacific Ocean La Niña events. Clim. Dyn. 42 (11–12), 3289–3311.

Hoell, A., Funk, C., Barlow, M., 2014. La Niña diversity and the forcing of Northwest Indian Ocean Rim teleconnections. Clim. Dyn. 42 (11–12), 3289–3311.

Huang, B., Thorne, P.W., Banzon, V.F., Boyer, T., Chepurin, G., Lawrimore, J.H., et al., 2017. Extended reconstructed sea surface temperature, version 5 (ERSSTv5): upgrades, validations, and intercomparisons. J. Clim. 30 (20), 8179–8205.

Husak, G.J., Michaelsen, J., Funk, C., 2007. Use of the gamma distribution to represent monthly rainfall in Africa for drought monitoring applications. Int. J. Climatol. 27 (7), 935–944.

Liebmann, B., Hoerling, M.P., Funk, C., Bladé, I., Dole, R.M., Allured, D., et al., 2014. Understanding recent Eastern Horn of Africa rainfall variability and change. J. Clim. 27 (23), 8630–8645.

Magadzire, T., Galu, G., Verdin, J., 2017. How climate forecasts strengthen food security. Bulletin 67 (Special Issue on Water).

McKee, T.B., Doesken, N.J., Kleist, J., 1993. The relationship of drought frequency and duration to time scales. In: Proceedings of the 8th Conference on Applied Climatology. Anaheim, CA. American Meteorological Society, Boston, MA.

Nicholson, S.E., 2015. Long-term variability of the East African 'short rains' and its links to large-scale factors. Int. J. Climatol. 35, 3979–3990. n/a-n/a.

Nicholson, S.E., 2017. Climate and climatic variability of rainfall over Eastern Africa. Rev. Geophys. 55, n/a-n/a.

Ropelewski, C.F., Halpert, M.S., 1987. Global and regional scale precipitation patterns associated with the El Niño/Southern Oscillation. Mon. Weather. Rev. 115 (8), 1606–1626.

Saji, N., Yamagata, T., 2003. Possible impacts of Indian Ocean dipole mode events on global climate. Clim. Res. 25 (2), 151–169.

Saji, N.H., Goswami, B.N., Vinayachandran, P.N., Yamagata, T., 1999. A dipole mode in the tropical Indian Ocean. Nature 401 (6751), 360–363.

WFP, 2017. Persistent drought in Somalia leads to major food security crisis, Multi-Agency Joint Alert, https://documents.wfp.org/stellent/groups/public/documents/ena/wfp290554.pdf.

CHAPTER 10

Practice—evaluating forecast skill

10.1 Introduction

Forecasts of climate, as well as climate's impacts on agriculture and water resources (e.g., soil moisture, streamflow, and groundwater), are crucial in implementing a successful drought early warning system (DEWS) to support decision-making. Forecasts intended to support a decision need to be timely and occur routinely, before the point in time when decisions need to be seriously considered. Forecasts also need to be available at the spatial and temporal scale most relevant to the decision context. In addition to the timeliness and appropriateness in terms of decision-making needs, the forecasts need to be skillful—or at least better than the alternatives. In order to assess the skill of forecasts, several methods have been suggested and are now widely used. The skill is generally expressed quantitatively in terms of a score—known as skill score. In general, the skill score provides a measurement of the performance of the forecasts during the past events. Typically, the number of events chosen for skill evaluation has to be sampled from several different years (generally, 30 years or so) to ensure that the skill score is indeed, to the extent possible, an accurate representation of the performance of the forecasts. Generally, skills of forecasts are also compared with the skill of "benchmark" forecasts. "Benchmark" forecasts can be "persistence forecasts," where it is assumed that the current state of a variable will simply persist into the future over the target period. "Climatological forecasts" may also be benchmarks, where it is assumed that the target forecast will be similar to long-term average conditions. Benchmarks could also be as simple as a "random" forecast, meaning the forecasts are made based on pure chance, with zero statistical or dynamical basis.

Forecast skill evaluation provides a rationale for the application of the forecasts for drought early warning and, hence, supports decision-making. Therefore forecast skill evaluation is an important component of implementing a DEWS. This chapter describes commonly used forecast evaluation methods, their usages, and their potential strengths and weaknesses, as well as providing examples of how forecast skill evaluation guides the application of such methods in DEWS.

Drought Early Warning and Forecasting
DOI: https://doi.org/10.1016/B978-0-12-814011-6.00010-5

Forecast skill evaluation methods can be broadly classified into two types: (1) deterministic forecast skill scores and (2) probabilistic forecast skill scores.

10.2 Deterministic forecast skill scores

Forecasts are inherently probabilistic due to the uncertainties associated with any future outcome. Typically, any climate or weather prediction system will generate multiple sets of future conditions. The models are initialized with current conditions and at each future time step, they simulate the variations in winds, clouds, temperatures, and precipitation. No two simulations will be the same. A set of such simulations is typically referred to as an ensemble. However, oftentimes for the sake of simplicity, an average (mean or median) of probabilistic forecast scenarios, often referred to as an "ensemble mean or median," is presented and communicated. Ensemble means can be calculated by giving equal weights to all the ensemble members or by varying weights based on a predetermined criterion (typically related to the skill of ensemble members). Ensemble mean forecast is also referred to as a deterministic forecast. The section below provides key details regarding frequently used deterministic forecast skill score methods.

Correlation: Correlation is among the most widely used deterministic forecasts skill score. Mainly, it indicates the linear relationship between the forecasts and observations. For seasonal scale climate forecasts, the time step at which correlation is calculated is often annual, and, in that case, correlation indicates the agreement in the interannual variability of the seasonal climate forecasts with the interannual variability of observations. The value of the correlation-based skill score varies from -1 to 1, with 0 showing no skill and 1 showing perfect skill. There are several methods to calculate correlation, with the most popular of the methods being Pearson's correlation and Spearman's rank correlation.

10.2.1 Pearson's correlation

Pearson's correlation is a measure of the linear relationship between forecasts and observations. It is calculated by dividing the covariance of forecasts and observations with the product of their standard deviations.

$$r = \frac{\sum_{i=1}^{i=n}(O_i - O_m)(F_i - F_m)}{\sqrt{\sum_{i=1}^{i=n}(O_i - O_m)^2}\sqrt{\sum_{i=1}^{i=n}(F_i - F_m)^2}}$$

where O_i and F_i are the ranks of observation and forecast, respectively at a time step i. n is the total number of time steps (i.e., number of events). O_m and F_m are mean of observations and forecasts, respectively, over the time step 0 to n.

Fig. 10.1, for example, shows the skill of December–February (DJF) precipitation and sea surface temperature (SST) forecasts from the North American Multimodel Ensemble (Kirtman et al., 2013) mean forecasts at 5 months before the season (Fig. 10.1A and B) and 1 month before the season (Fig. 10.1C and D). The skill is shown in terms of correlation with observed precipitation and SST, respectively. In this case the skill is calculated at the native spatial resolution of the forecasts, which is 1-degree latitude × 1-degree longitude. In cases where the observations are available at finer spatial resolution than forecasts, then the first step before skill calculation would be to spatially aggregate the observations to match the spatial resolution of the forecasts.

Fig. 10.1 highlights a few of the key points regarding forecasts skill.

10.2.1.1 Higher skill at lower lead

By comparing Fig. 10.1A with C, and Fig. 10.1B with D, it can be seen that, in general, the forecasts skill drops as the lead time, which is the time between when the forecasts are made and the target forecast period (i.e., DJF season in the case of Fig. 10.1). The drop in skill mainly has to do with the inherent uncertainty in the forecasts, which tends to increase as the time since the forecasts initialization increases, following Lorenz's chaos theory.

10.2.1.2 Lower skill in forecasting terrestrial precipitation

Fig. 10.1A and C shows that some of the highest precipitation forecasts skill regions are over the ocean. The skill over the terrestrial regions are generally lower and often negligible. This general lack of precipitation forecasts skill has to do with the spatial distance (leading to greater uncertainty) between the terrestrial regions and the oceans, which are often the source of the moisture and lack of adequate representations of mountains and land cover, which affects the precipitation formulation process, and in some cases, the source of the moisture, which can be terrestrial.

10.2.2 Spearman's rank correlation

Pearson's correlation method, described previously, is a widely used method for calculating forecast skill scores; however, it assumes that a linear relationship exists between both variables and that they are normally distributed. This method of correlation calculation is also sensitive to the

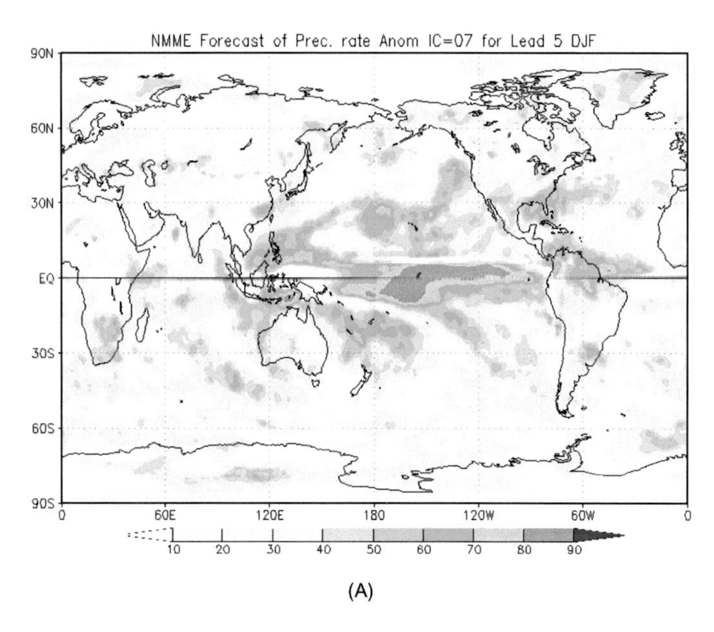

(A)

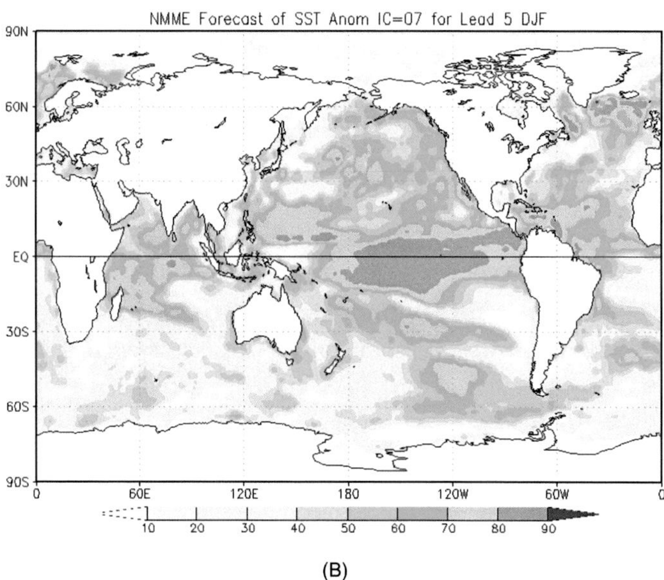

(B)

Figure 10.1 Forecast skill, in terms of correlation, of North American Multimodel Ensemble mean DJF precipitation forecasts made (A) 5 months and (C) 1 month before the season. Parts (B) and (C) are the same but for DJF SST forecasts. These skill maps are provided along with the operational forecasts by the U.S. CPC https://www.cpc.ncep.noaa.gov/products/NMME/. *CPC*, Climate Prediction Center; *DJF*, December–February; *SST*, sea surface temperature.

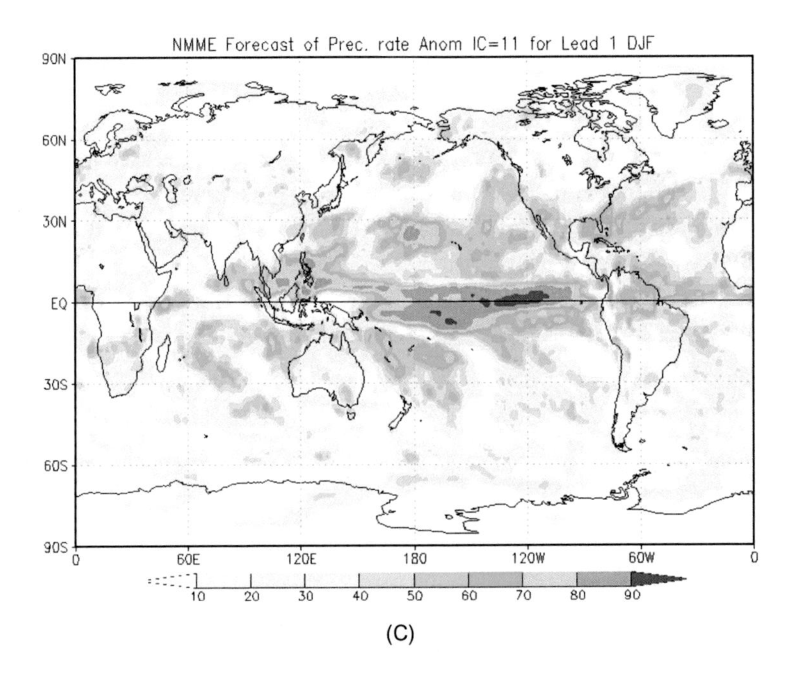

(C)

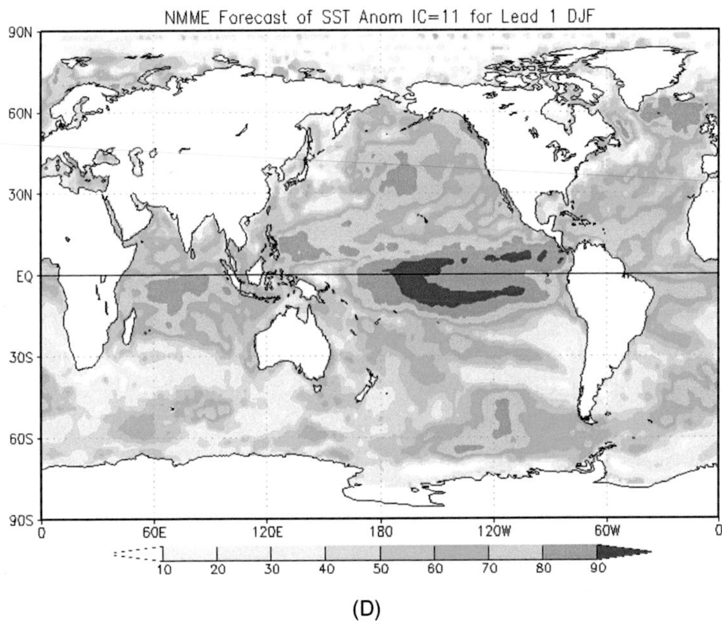

(D)

Figure 10.1 (Continued)

extreme values. Alternatively, the Spearman's rank correlation method is calculated using the rank of forecasts and observations rather than absolute values; hence, it is less sensitive to extreme values. The Spearman's rank method examines a monotonic relationship between forecasts and observations; in other words, it checks if the forecasts increases or decreases as observation increases or decreases. The quantity of the increase or decrease in forecasts and observations does not have to be proportionate. The method to calculate Spearman's rank correlation is similar to the method of calculating Pearson's correlation, with the main difference being that the Spearman's rank correlation is calculated using the ranks of the forecasts and observations.

$$r = \frac{\sum_{i=1}^{i=n} (\mathrm{Rank}O_i - \mathrm{Rank}O_m)(\mathrm{Rank}F_i - \mathrm{Rank}F_m)}{\sqrt{\sum_{i=1}^{i=n} (\mathrm{Rank}O_i - \mathrm{Rank}O_m)^2}\sqrt{\sum_{i=1}^{i=n} (\mathrm{Rank}F_i - \mathrm{Rank}F_m)^2}}$$

where $\mathrm{Rank}O_i$ and $\mathrm{Rank}F_i$ are the ranks of observation and forecast, respectively, at a time step i. n is the total number of time steps (i.e., number of events). $\mathrm{Rank}O_m$ and $\mathrm{Rank}F_m$ are the mean of rank of observations and forecasts, respectively, over the 0 to n time steps.

10.2.3 Equitable threat score (Gilbert skill score)

During the examination of the relationships between forecasts and observations, for decision-making purposes, it is often useful to know how well given forecasts can distinguish between two categories of events. For example, the categories can be "no rain" or "rain," or, more specific to seasonal drought forecasting, the categories can be "below normal" or "not below normal." The equitable threat score (ETS) is a widely used skill score that measures how well the forecasts identify a certain category of event relative to a random forecast. More specifically, this skill score estimates the fraction of "hits" after accounting for the number of "hits" that can be obtained purely due to random chance, where "hits" signify the cases where a certain category of an event is observed and is correctly forecasted. "Misses" signify the cases where a certain category of an event happened but was not forecasted, meaning forecasts missed the event.

The ETS value varies from −1/3 to 1 where 0 indicates no skill and 1 indicates perfect forecast for a given category.

$$\text{ETS} = \frac{\text{hits} - \text{hits}_{\text{random}}}{\text{hits} + \text{misses} + \text{falsealarms} - \text{hits}_{\text{random}}}$$

where

$$\text{hits}_{\text{random}} = \frac{(\text{hits} + \text{misses})(\text{hits} + \text{falsealarms})}{\text{total}}$$

hits is the number of times the observation event category is correctly identified (if observation category is "yes," the forecast category is also "yes"); misses is the number of times the observation event category is not identified (if observation category is "yes," the forecast category is "no"); and falsealarms is the number of times the observation event category is not misidentified (if observation category is "no," the forecast category is "yes").

In order for ETS to be useful for decision-making applications, it is important to identify dichotomous categories of events that are directly relevant to the target decisions. For example, if decisions are drought related, the focus category might be "below a certain percentile threshold" (such as "below normal" or "below 33 percentile"). For interested readers, Shukla et al. (2016) shows the ETS of NMME precipitation and temperature forecasts in identifying above (>67 percentile) and below (<33 percentile) normal events in the Greater Horn of Africa.

10.3 Probabilistic forecast skill scores

10.3.1 Brier skill score

The Brier skill score (BSS) is commonly used to evaluate the quality and reliability of probabilistic forecasts. This score indicates how well the forecasted probability of given events corresponds with the observed frequency. BSS is the mean squared error of the forecast probability (Wilks, 2011).

$$\text{BS} = \frac{1}{N}\sum_{k=1}^{N}(F_k - O_k)^2$$

where BS is Brier skill, N is the number of forecast event pairs, F_k is forecast probability of a given event (can vary from 0 to 1), and O_k can be 0 or 1 depending on whether the event occurred in the category *k* or not. The value of this score varies from 0 to 1, with 0 being the score for perfect forecasts. The closer the forecast probability is to the observed probability (of 0 or 1), the lower and better the score will be. For example, if an event was observed, the observation probability will be 1, and if the forecast probability was closer to 1, the BS will be lower and hence the forecast performance will be higher.

The BSS is calculated by comparing BS of a given forecast with the BS of a reference forecast, such as climatology.

$$\text{BSS} = 1 - \frac{\text{BS}}{\text{BS}_{\text{ref}}}$$

where BS_{ref} is Brier skill score of a reference forecast (often climatology), BSS of 1 will indicate perfect forecasts, whereas BSS of 0 will indicate forecasts with no skill.

The BSS can be used to track skill of forecasts in near real time. For example, Fig. 10.2 shows the BSSs of the Climate Prediction Center's seasonal forecasts of precipitation and temperature from the January 2010 to December 2018 period. As can be seen in this figure, the skill of temperature is generally higher than the skill of precipitation, and that skill varies with time. Note that skill score values shown in this figure are values averaged over the Conterminous United States and can be higher for specific regions and seasons.

10.3.2 Ranked probability skill score

The ranked probability skill score (RPSS) is similar to the BSS in that it is a probabilistic skill score that also indicates how accurate the forecasts probabilities are relative to reference forecasts, such as climatological forecasts. The main difference between the two is the number of categories of events. RPSS events are divided into multiple categories such as tercile categories, quartiles, or sorted into an even larger number of categories. Forecasted probabilities for each category of the events are compared with the observed frequency, and forecasts are rewarded for assigning higher probability closest to the category in which observation lies. For example, if during a given event, the observation falls into the above-normal category, the forecasts that assigned higher probability to that category will be considered more skillful than the forecasts that assigned the lowest probability to that category. In other words,

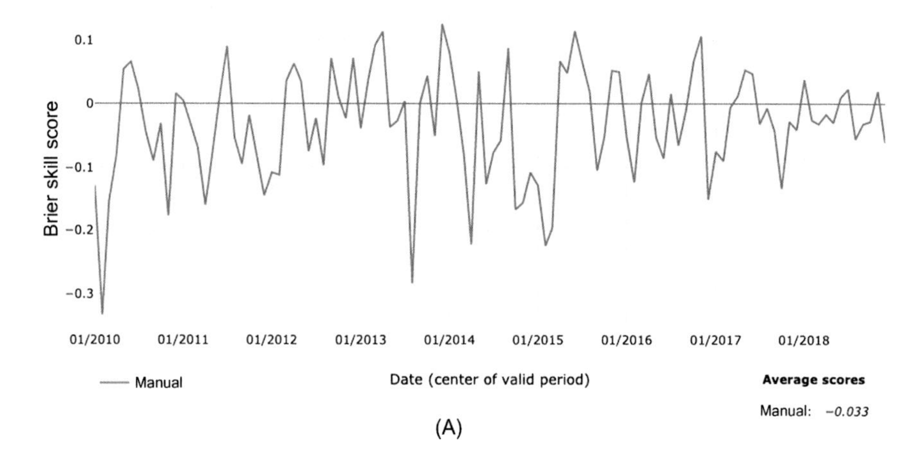

(A)

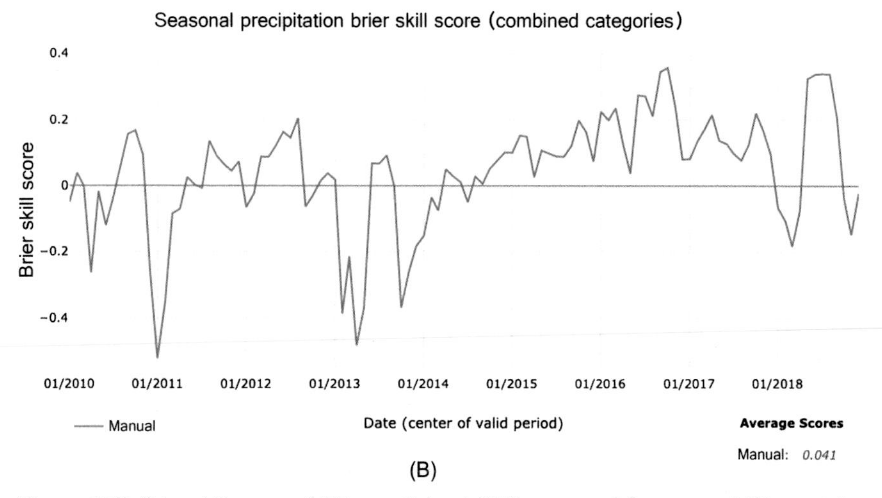

(B)

Figure 10.2 Brier skill score of 0.5 month-lead CPC's seasonal forecasts of (A) precipitation and (B) temperature, over the Conterminous United States over the period of January 2010 through December 2018. The plots were generated using CPC Verification Web Tool (https://vwt.ncep.noaa.gov/index.php?page = chart). *CPC*, Climate Prediction Center.

the RPSS rewards a forecast for the number of ensemble members that fall within the observed category. RPSS varies from 0 to 1. Scores greater than 0 indicates that the given forecast is more skillful than the reference forecasts (i.e., climatological forecasts), and a score of 1 indicates a perfect forecast.

The ranked probability score (RPS) is calculated as follows:

$$\mathrm{RPS} = \frac{1}{(N-1)}\sum_{n=1}^{N}\left[\sum_{k=1}^{n} F_k - \sum_{k=1}^{n} O_k\right]^2$$

where RPS is ranked probability score, N is the number of forecast event pairs, F_k is forecast probability for category k, and O_k can be 0 or 1 depending on if the event occurred in the category k or not. The value of score varies from 0 to 1, with 0 being the score for perfect forecasts.

The RPSS of a forecast is calculated by comparing the RPS of the forecast relative to the RPS of a reference forecast such as climatological forecast.

$$\mathrm{RPSS} = 1 - \frac{\mathrm{RPS}_m}{\mathrm{RPS}_{\mathrm{ref}}}$$

where RPS_m is the average ranked probability score of the given forecast (over a given time period) and $\mathrm{RPS}_{\mathrm{ref}}$ is the average ranked probability score of the reference forecast. The value of the RPSS varies from $-\infty$ to 1, where RPSS of 1 indicates a perfect forecast.

The RPSS is one of the skill scores that the International Research Institute uses to examine the skill (Goddard et al., 2003) of their precipitation, temperature, and SST forecasts. Fig. 10.3 shows the skill of 0.5 month-lead precipitation and temperature forecasts for all seasons over the globe. As in Figs. 10.1 and 10.2, this figure also emphasizes higher skill in the case of temperature relative to precipitation. The figure also shows the skill to be typically higher in the tropics, especially in the case of temperature.

10.3.3 Reliability diagram

Another widely used method to examine the skill of probabilistic forecast is the reliability diagram (Hartmann et al., 2002). As the name implies, this method measures the reliability of forecasts, which, simply put, means how accurate forecast probabilities are. A reliability diagram is made by comparing the forecast probabilities relative to the observed frequencies. This diagram is of direct value for decision-making purposes, as it tells the decision-makers how often a given forecast probability of a given category was actually realized in the observations. A perfectly reliable ensemble forecast would fall on 1:1 line. For example, a perfectly reliable forecast ensemble provides X% of probability to events those events will happen

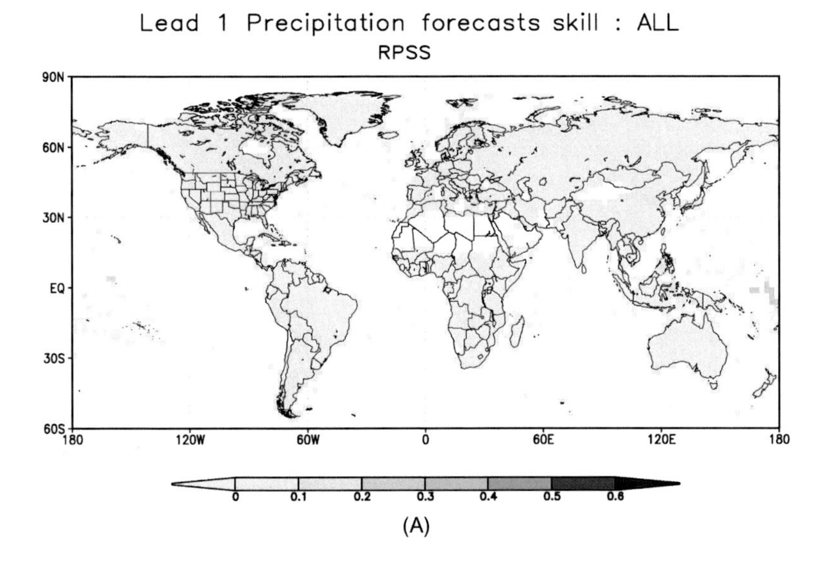

(A)

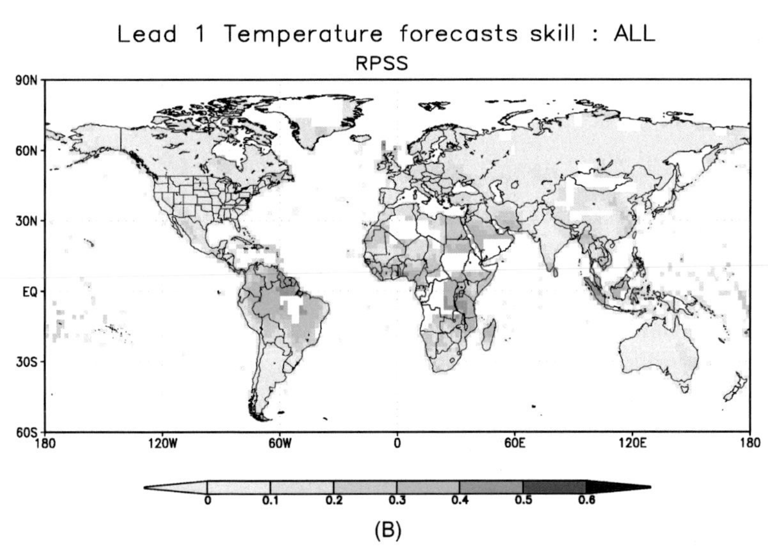

(B)

Figure 10.3 Ranked probability skill score of lead 1 IRI's seasonal forecasts of (A) precipitation and (B) temperature, over the globe and for all the seasons. The plots were obtained from https://iri.columbia.edu/our-expertise/climate/forecasts/verification/. *IRI*, International Research Institute.

about $X\%$ of times. The reliability diagram is prepared by partitioning forecast probabilities into multiple categories (e.g., probabilities of 0–1 can be divided into 10 categories of 0.1 probability size bin.) For each bin

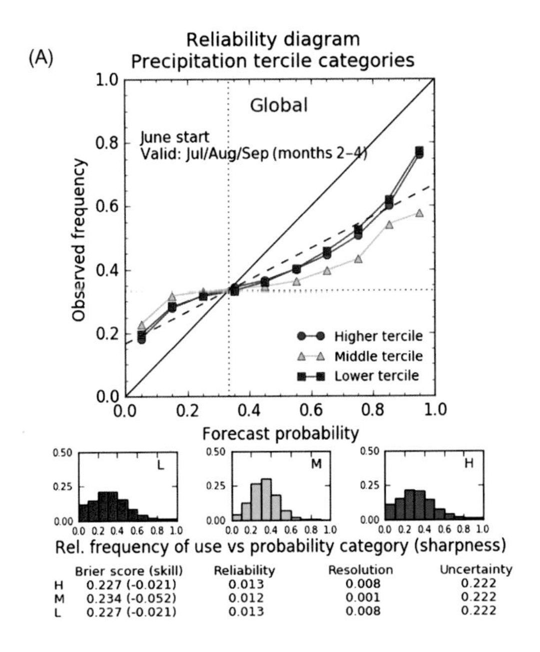

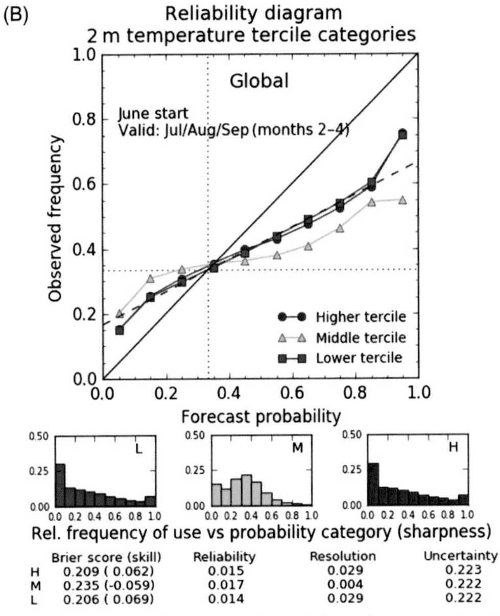

Figure 10.4 Reliability diagram of the UK Met Office forecasts lead 1 (A) JJA precipitation, (B) JJA temperature, (C) DJF precipitation, and (D) DJF temperature over the globe. The plots were obtained from the UK met office (https://www.metoffice.gov.uk/research/climate/seasonal-to-decadal/gpc-outlooks/glob-seas-prob-skill). *DJF*, December–February.

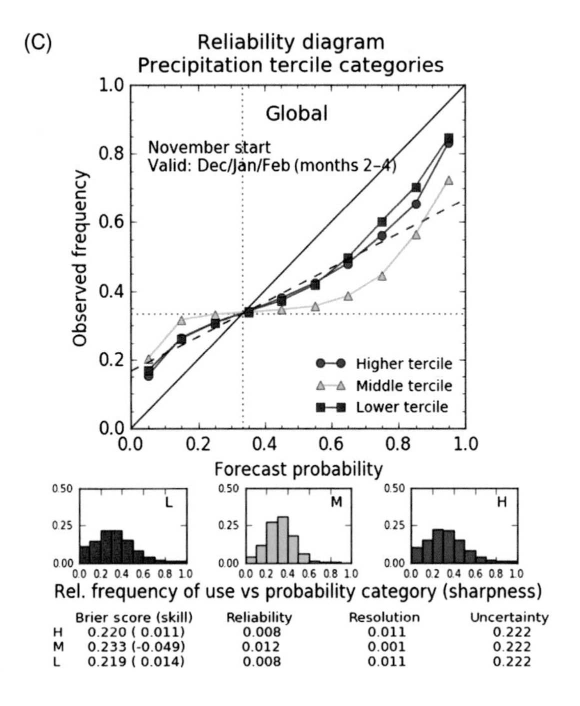

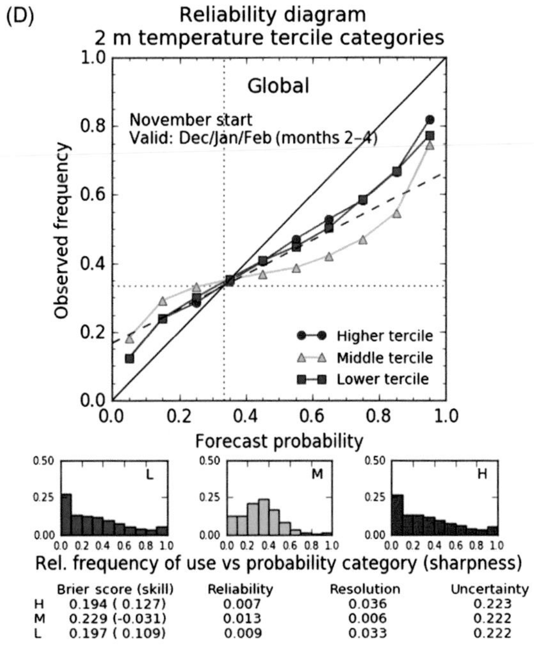

Figure 10.4 (Continued)

the observed frequency is calculated to examine the correspondence of forecast probability with observed frequency.

The hit rate for each of the bins is calculated as follows:

$$\text{Hit}_{\text{rate}} = \frac{O_n}{O_n + NO_n}$$

where O_n is the number of observed instances in the given probability bin and NO_n is the number of instances not observed in the given probability bin.

Reliability diagrams can be made for a given location or for a large region, when forecasts over different pixels in a given probability bin are pooled together. For example, Fig. 10.4 shows the reliability diagram over the globe of the June–July–August (JJA) precipitation and temperature forecasts (Fig. 10.4A and B) and DJF precipitation and temperature forecasts by the UK Met Office's seasonal forecasts (MacLachlan et al., 2015). Reliability is shown for each of the tercile categories.

The difference between the reliability curve of the tercile forecasts and the 1:1 line depicts the reliability of the forecasts. For example, in general, these figures show that these forecasts tend to overestimate the probability confidence. When forecast probability values are high, the reliability curve is below the 1:1 line, which indicates that observation frequencies are typically smaller than forecast probability. These reliability diagrams also indicate that, in general, forecasts for above- and below-normal categories are more reliable than the normal category, as the reliability cures for above- and below-normal categories are close to the 1:1 line.

10.4 Summary

Weather and climate forecasts are crucially important for an effective DEWS. The skill of a DEWS directly depends on the skill of weather and climate forecasts. This chapter presents a few of the widely used metrics used for calculating skill of weather and climate forecasts. Those metrics can be mainly classified into deterministic and probabilistic metric scores. The chapter also highlights typical tendencies of how skill varies with the lead time and the areas and variables of most skill.

References

Goddard, L., Barnston, A.G., Mason, S.J., 2003. Evaluation of the IRI'S "Net Assessment" seasonal climate forecasts: 1997–2001. Bull. Am. Meteorol. Soc. 84, 1761–1782. Available from: https://doi.org/10.1175/BAMS-84-12-1761.

Hartmann, H.C., Pagano, T.C., Sorooshian, S., Bales, R., 2002. Confidence builders. Bull. Am. Meteorol. Soc. 83, 683–698. Available from: https://doi.org/10.1175/1520-0477(2002)083 < 0683:CBESCF > 2.3.CO;2.

Kirtman, B.P., et al., 2013. The North American Multimodel Ensemble: phase-1 seasonal-to-interannual prediction; phase-2 toward developing intraseasonal prediction. Bull. Am. Meteorol. Soc. 95, 585–601. Available from: https://doi.org/10.1175/BAMS-D-12-00050.1.

MacLachlan, C., et al., 2015. Global seasonal forecast system version 5 (GloSea5): a high-resolution seasonal forecast system. Q. J. R. Meteorol. Soc. 141, 1072–1084. Available from: https://doi.org/10.1002/qj.2396.

Shukla, S., Roberts, J., Hoell, A., Funk, C.C., Robertson, F., Kirtman, B., 2016. Assessing North American multimodel ensemble (NMME) seasonal forecast skill to assist in the early warning of anomalous hydrometeorological events over East Africa. Clim. Dyn. . Available from: https://doi.org/10.1007/s00382-016-3296-z.

Wilks, D.S., 2011. Statistical Methods in the Atmospheric Sciences. Academic Press, 698 pp.

CHAPTER 11

Practice—integrating observations and climate forecasts

11.1 Approach

Because droughts are slow onset disasters, predictive skill can come from monitoring systems and climate and weather forecasts. For example, simply observing rainfall deficits can be an important indicator of future crop failures of water supply crises. To gain even more lead time, however, it is useful to leverage the skill provided by weather and climate forecasts. Combining these information sources in integrated monitoring and prediction systems can seamlessly incorporate information from the ocean, land, and atmosphere (Fig. 9.2). But to routinely achieve this goal, these information sources must be made "interoperable." Typically, coarse resolution forecast information needs to be rescaled to align with higher-resolution observation data. Fig. 11.1 provides an overview of the basic process of merging observations with climate forecasts and the main steps involved. Climate forecasts are typically available at a coarser spatial scale than the spatial scale at which impact models operate, and the scale at which drought decisions are frequently made. In addition, climate forecasts are typically biased relative to observational data sets. This mismatch in scale and inherent bias makes it imperative to bias-correct and downscale climate forecasts before using them to drive impact models.

In the following section a brief description of a few of the common bias-correction and downscaling methods is provided. The initial state—or the state of soil moisture or snow, which represent current hydrologic conditions—is obtained by running the impact model using observed atmospheric forcings. These observed atmospheric forcings can be based on in situ observations from weather stations, or satellites, or reanalysis, or all of the above. The length of the model simulation to generate the best estimate of initial conditions varies depending on the model and the target initial state.

Post bias-correction, starting from the initial conditions, impact forecasts are made by driving the impact models using bias-corrected forecasts.

Drought Early Warning and Forecasting
DOI: https://doi.org/10.1016/B978-0-12-814011-6.00011-7

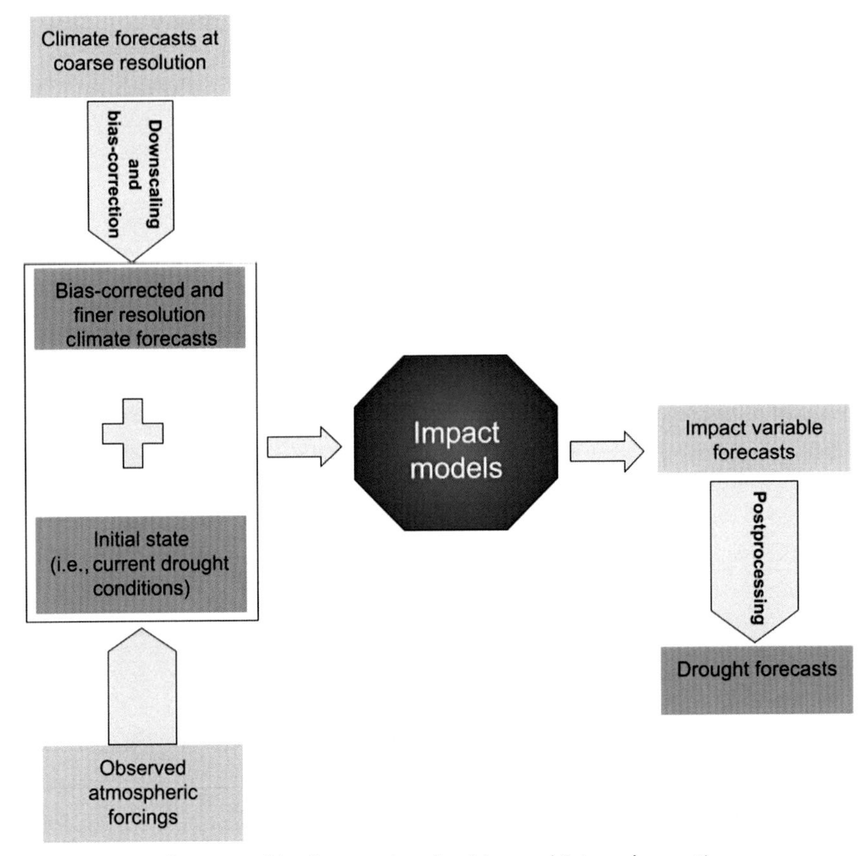

Figure 11.1 Schematic of basic steps involved in combining observations.

The length of the forecast period depends on the period for which climate forecasts are available, for target decisions as well as the skill of the impact forecasts. Often the skill of climate forecasts is minimal beyond the first few months. There are times, however, such as during the middle of a peak precipitation season, when outlooks may be very skillful for many months. A poor start to a rainy season, identified by observations, combined with a 1-to 2-month climate forecast indicating continued dry and warm weather, can often predict water deficits that may persist for up to 9 months, when the seasonal cycle brings next year's rains and an opportunity for recovery. Integrated observation–climate–impact models are ideally suited to providing outlooks within this important window of forecast opportunity. A key component of such systems, however, is the downscaling of climate forecasts, discussed here.

Once the impact forecasts are generated, the next important step is to postprocess them to convert them into drought indicator forecasts. For example, soil moisture forecasts are typically converted into soil moisture percentiles. This places potential drought in historic context.

11.2 Bias-correction and downscaling methods

As mentioned earlier, bias-correction and downscaling are crucial steps in providing impact forecasts by integrating observations with climate forecasts. Several statistical methods as well as dynamical downscaling methods exist. Statistical downscaling is widely used, however, as it is numerically less expensive. The benefits provided by the substantially higher additional computational time due to dynamical downscaling are not always apparent. A few of the widely used methods for statistical downscaling are described in the following sections. Most of the downscaling methods described here were originally proposed for downscaling climate change projections; however, in different cases, they have used to downscale seasonal scale climate forecasts as well.

11.2.1 Bias-correction and spatial downscaling method

The bias-correction and spatial downscaling (BCSD) method, as the name implies, corrects bias in coarse-resolution climate model outputs (applicable for other types of coarse-resolution model outputs as well) and then downscales the bias-corrected outputs. The method is primarily based on the widely used quantile mapping approach and was proposed for use in deriving climate forcings to drive hydrologic models for seasonal scale hydrologic forecasting applications (Wood et al., 2002), as well as for long-term hydrologic projections (Wood et al., 2004). The process of implementing BCSD on seasonal-scale climate forecasts and long-term climate projections is generally the same, with the only main difference being that the assumption of nonstationarity in the climate is more appropriate for seasonal-climate forecasts than at the climate projection scale. The two main steps in the process, as mentioned previously, are (1) bias-correction and (2) spatial downscaling.

11.2.1.1 Bias-correction

Bias-correction is performed by using a quantile mapping approach. In this approach a bias-corrected value for a target forecast value (or projections) is obtained by first converting the target forecast value into nonexceedance probability using the "forecast" climatology (often referred to as hindcasts). Once the target value is converted into nonexceedance probability, a bias-corrected value is obtained by using the nonexceedance probability to query the corresponding value from the "observed" climatology. Typically, the longest overlapping period between the historical "observations" and "forecasts" is used for each climatology. Also, for "forecast" climatology, all ensemble members of the hindcasts are used to form one climatology, as the assumption is that different model ensemble members come from the same climatology. For example, a model with 10 ensemble members over the 1981–2010 climatological period will have 300 members in the climatology: 30—number of years—multiplied by 10—number of ensemble members, assuming the climatology is made of monthly/seasonal values. For daily values the number of events will increase by the length of the moving window period (often a 15-day window, centered on the target forecast day). The choice of the climatological period—both in terms of the number of years to include and in the case of submonthly forecasts (such as daily forecasts)—is an important one. The climatology should ideally include the largest possible sample of the events that are most representative of the climate of the target forecasts. For example, when bias-correcting daily forecasts, it is a common practice to include all the forecasts for the days within a 15-day window of the target day (e.g., January 1–15, to bias-correct a target forecast of January 8).

Similarly, the important choice to be made is the type of distribution to use. Often, the choice is between using empirical distributions versus theoretical distributions, and if theoretical distributions are used, then the choice of which theoretical distribution to use must be made.

11.2.1.2 Spatial downscaling

Once the bias-correction is done, the next step is downscaling. Typically, after bias-correction, the bias-corrected values are converted into anomalies (additive anomalies in the case of continuous variables such as temperature and multiplicative anomalies in the case of noncontinuous variables such as precipitation). The anomalies are then interpolated to the grid of the fine-scale observations. For interpolation, there are several available methods such as bilinear and bicubic interpolation. Once the interpolation is done, the anomalies are applied onto the fine-scale "observed" climatology to get bias-corrected values at the fine-scale spatial resolution. Depending on the cases following the spatial downscaling, temporal downscaling may be required.

11.2.2 Constructed analog method

The constructed analog (CA) method, proposed in Hidalgo et al. (2008), has been primarily used for downscaling climate change projections. In this method a "best" set of analogs of the future climate events, taken from an archive of the past observed climate, is combined (typically weighted average) to provide downscaled versions of the future climates. Predictors used in this method are typically synoptic scale fields (such as precipitation, temperature, and geopotential heights). Future projections of those synoptic-scale fields are compared with a library of past observed synoptic scale fields (from a reanalysis or observational based data set, etc.) for the same time of the year (similar to the BCSD method), and "best" analogs are selected (typically the top 30). Based on the similarity of each analogue with the future conditions, they are assigned weights, which are used to provide a weighted mean of the analogs. The weighted mean is used as "downscaled" climate projections. The basic assumption in the CA method is dynamically sound, and it also takes into account the regional complexity and spatial coherence better than the BCSD method and can actually enhance the skill (i.e., performance in capturing interannual variability) of seasonal-scale climate forecasts, as demonstrated in (Shukla et al., 2014). However, this method does not account for the bias in the climate model outputs; hence, a bias-correction step is needed either before or after applying the CA method (Maurer et al., 2010; Abatzoglou and Brown, 2012). The implementation of the CA method can be summarized by the following equations:

$$Z_{coarse} = Z_{analogs} A_{analogs} \tag{11.1}$$

where $Z_{analogs}$ are the best set of analogs from the library of the coarse resolution patterns of synoptic fields. Z_{coarse} is the target pattern and $A_{analouges}$ is the least-square estimates of the regression coefficients that indicate the contribution of each of the $Z_{analogs}$ to constructing Z_{coarse}

$A_{analogs}$ is obtained from Eq. (11.3):

$$A_{analogs} = \left[\left(Z'_{analogs} Z_{analogs} \right)^{-1} Z'_{analogs} \right] Z_{coarse} \tag{11.2}$$

where $Z'_{analogs}$ denotes the matrix transpose. Finally using the linear regression coefficients $A_{analogs}$ and fine-scale analogs $P_{analogs}$ at the same time steps as $Z_{analogs}$, downscaled synoptic fields $P_{downscaled}$ can be estimated following Eq. (11.4):

$$P_{downscaled} = P_{analogs} A_{analogs} \tag{11.3}$$

11.2.3 Multivariate Adaptive Constructed Analogs

The Multivariate Adaptive Constructed Analogs (MACA) method (Abatzoglou and Brown, 2012) was proposed to address the limitations in the CA method. First, the MACA method uses the daily BCSD method to correct the bias in the general circulation model outputs before the application of the CA method, which the CA method does not do. Next, an Epoch adjustment is done to account for another limitation of the CA method in the cases when no analogs of the future climates exist in the past. Epoch adjustment removes the differences in the mean of the future and the past climate. Next, the CA method is applied on the bias-corrected and Epoch-adjusted fields. The MACA method also performs a multivariate search for analogs—for example, based on temperature max, min, and dew point, rather than one field only. This step helps ensure the dynamical consistency among different related variables. After the CA method the resulting downscaled product again goes through Epoch adjustment based on the differences calculated in the previous step. Finally, after the second Epoch adjustment, the resulting downscaled product is again bias-corrected following the BCSD method. Downscaled climate projections using MACA can be found at https://climate.northwestknowledge.net/MACA/index.php.

11.2.4 Bayesian merging

Bayesian merging of climate model outputs to drive impact models such as hydrologic models is based on the Bayes' theorem. Bayes' theorem provides the probability of an event based on the knowledge of conditions that are related to that event. The application of Bayes' technique for merging climate model outputs to drive hydrologic models is presented in (Luo et al., 2007; Luo and Wood, 2008). The implementation of Bayes' theorem for merging climate model outputs is carried out by the following equation:

$$p(\theta|y) = \frac{p(\theta|y)p(y|\theta)}{p(y)} \tag{11.4}$$

where $p(\theta)$ is the assumed prior distribution of θ. Often the climatological distribution is assumed as the prior distribution. $p(y|\theta)$ is the likelihood function that describes the relationship (estimated beforehand) between the probability of θ given the prior knowledge of y. $p(\theta y)$ is the posterior distribution that provides the updated probability of θ due to the prior knowledge of y.

The primary benefit of the Bayes' merging approach, as described in Luo and Wood (2008), is that it corrects the forecast probability based on the prior knowledge of the performance of the historical forecasts. It also corrects for bias and downscales the forecast, through the likelihood function, which relates forecasts at coarse spatial resolution with observations at fine spatial resolution.

11.3 An example: The NASA Hydrological and Forecast Analysis System

The National Aeronautics and Space Agency (NASA) Hydrological and Forecast Analysis System (NHyFAS) provides forecasts of soil moisture over the next 6 months, for Africa and the Middle East regions, which are prone to water and food insecurity events. The maps from this system are provided near the mid-month via https://lis.gsfc.nasa.gov/projects/nhyfas. Fig. 11.2 provides an overview of the NHyFAS framework.

This system is built upon NASA's Land Information System (LIS, https://lis.gsfc.nasa.gov/), which allows for a high-performance terrestrial hydrology modeling and data assimilation using multiple land surface models. The Climate Hazards InfraRed Precipitation with Station (Funk et al., 2015) and NASA's reanalysis data set MERRA-2 (Gelaro et al., 2017) are used as observed atmospheric forcings to drive the LIS

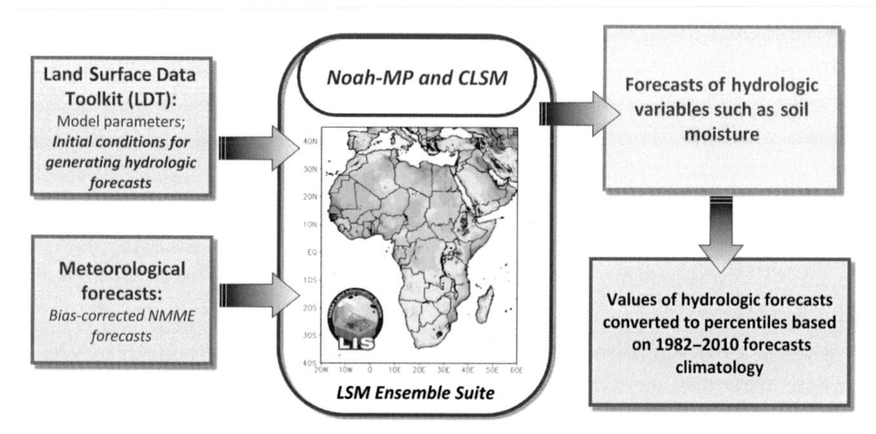

Figure 11.2 Overview of the NHyFAS. *NHyFAS*, NASA Hydrological and Forecast Analysis System. *Adopted from Arsenault, K. R., Shukla, S., Hazra, A., Getirana, A., Mcnally, A., Kumar, S. V., ... Verdin, J. P. (2020). The NASA hydrological forecast system for food and water security applications.* Bulletin of the American Meteorological Society. *https://doi.org/10.1175/bams-d-18-0264.1.*

framework to (1) generate long-term climatology of hydrologic variables such as soil moisture used for drought forecasting and (2) provide an estimate of initial conditions at the time of the hydrologic forecast initialization. The NHyFAS uses North American Multimodel Ensemble (NMME, Kirtman et al., 2013) climate forecasts to provide drought forecasts, after bias-correction and downscaling. The BCSD method is used to correct the bias in climate forecasts relative to the observed atmospheric forcings mentioned previously. Since the NMME forecasts are typically available by the 10th of the month, the hydrologic forecasts derived using them are made available around the 15th of the month. These hydrologic forecasts (i.e., forecasts of soil moisture) are then converted into soil moisture percentiles, which are an indicator of agricultural drought, thereby informing drought forecasts and early warning.

Primary end users of this drought forecasting system include USAID's Famine Early Warning Systems Network (FEWS NET) (https://fews.net/). The FEWS NET provides early warning of food insecurity outlooks in Africa (among other regions such as Central America and Central Asia). Climate-driven extremes such as drought and flood contribute to food insecurity in this region; hence, the NHyFAS supports the early warning efforts of FEWS NET by routinely providing updates of expected drought (and flood) conditions throughout Africa.

11.4 Summary

Climate forecasting is an important tool for supporting drought forecasting. The use of climate forecasts to drive impact models (such as hydrologic models or crop models) to provide impact forecasts is done to (1) provide forecasts of variables that are directly relevant to drought decision-making and (2) allow for the integration of observations with climate forecasts. This process can further increase the skill and value of the impact forecasts, because the integration of observations can add greatly to the overall skill. The integration of climate forecasts with observations, however, requires bias-correction and downscaling due to mismatches in the scale at which climate forecasts are available and the scale at which impact models are run, as well as the inherent bias in the climate forecasts relative to observations. This chapter has described a few of the common statistical downscaling methods and lists their potential strengths. The chapter also provides an example of the integration of observation and climate forecasts for providing drought forecasts over Africa and the Middle East.

References

Abatzoglou, J.T., Brown, T.J., 2012. A comparison of statistical downscaling methods suited for wildfire applications. Int. J. Climatol. 32, 772–780. Available from: https://doi.org/10.1002/joc.2312.

Funk, C., et al., 2015. The climate hazards infrared precipitation with stations—a new environmental record for monitoring extremes. Sci. Data 2, 1–21. Available from: https://doi.org/10.1038/sdata.2015.66.

Gelaro, R., et al., 2017. The Modern-Era Retrospective Analysis for Research and Applications, Version 2 (MERRA-2). J. Clim. 30, 5419–5454. Available from: https://doi.org/10.1175/JCLI-D-16-0758.1.

Hidalgo, H.G., Dettinger, M.D., Cayan, D.R., 2008. *Downscaling with Constructed Analogues: Daily precipitation and temperature Fields Over the United States PIER Energy-Related Environmental Research Report CEC-500-2007-123* (Sacramento, CA: California Energy Commission).

Kirtman, B.P., et al., 2013. The North American Multimodel Ensemble: Phase-1 seasonal-to-interannual prediction; Phase-2 toward developing intraseasonal prediction. Bull. Am. Meteorol. Soc. 95, 585–601. Available from: https://doi.org/10.1175/BAMS-D-12-00050.1.

Luo, L., Wood, E.F., 2008. Use of Bayesian Merging Techniques in a multimodel seasonal hydrologic ensemble prediction system for the Eastern United States. J. Hydrometeorol. 9, 866–884. Available from: https://doi.org/10.1175/2008JHM980.1.

Luo, L., Wood, E.F., Pan, M., 2007. Bayesian merging of multiple climate model forecasts for seasonal hydrological predictions. J. Geophys. Res. Atmos. 112. Available from: https://doi.org/10.1029/2006JD007655.

Maurer, E.P., Hidalgo, H.G., Das, T., Dettinger, M.D., Cayan, D.R., 2010. The utility of daily large-scale climate data in the assessment of climate change impacts on daily streamflow in California. Hydrol. Earth Syst. Sci. 14, 1125–1138. Available from: https://doi.org/10.5194/hess-14-1125-2010.

Shukla, S., Funk, C., Hoell, A., 2014. Using constructed analogs to improve the skill of National Multi-Model Ensemble March-April-May precipitation forecasts in equatorial East Africa. Environ. Res. Lett. 9, 094009. Available from: https://doi.org/10.1088/1748-9326/9/9/094009.

Wood, A.W., Maurer, E.P., Kumar, A., Lettenmaier, D.P., 2002. Long-range experimental hydrologic forecasting for the eastern United States. J. Geophys. Res. Atmos. 107, ACL 6-1–ACL 6-15. Available from: https://doi.org/10.1029/2001JD000659.

Wood, A.W., Leung, L.R., Sridhar, V., Lettenmaier, D.P., 2004. Hydrologic implications of dynamical and statistical approaches to downscaling climate model outputs. Clim. Change 62, 189–216. Available from: https://doi.org/10.1023/B:CLIM.0000013685.99609.9e.

CHAPTER 12

Practice—actionable information and decision-making networks

While the origin of the concept is unclear, it is common to place decision-making in the context of the data–information–knowledge–wisdom hierarchy (Fig. 12.1). In the experience of a person, "data" might be the raw sensations experienced via touch, taste, hearing, vision, and smell. In and of themselves, these stimuli provide little value. When interpreted by the brain, these sensations become meaningful. Noise becomes language, or an identifiable sound, such as laughter or the roar of a jet engine. Through this process of distillation, interpretation, and contextualization, data becomes information. Data as information is placed into context, used to answer questions, and given specific relevance or meaning. In the context of a drought early warning system (DEWS), data might refer to the recorded mercury level in a weather station thermometer, or the thermal infrared irradiance observed by a weather satellite. Such data begins its journey toward information through interpretation as specific geophysical variables. The development of timely, accurate, and robust flows of such information is a necessary but insufficient aspect of an effective DEWS. Petabytes of information do not an effective early warning system make. The same caveat holds for online portals and websites. Providing websites with vast content is not synonymous with an effective DEWS. Effectiveness arises from (1) the translation of drought information into reasonably accurate and timely estimates of drought impacts and (2) the combination of such information with effective mitigation and

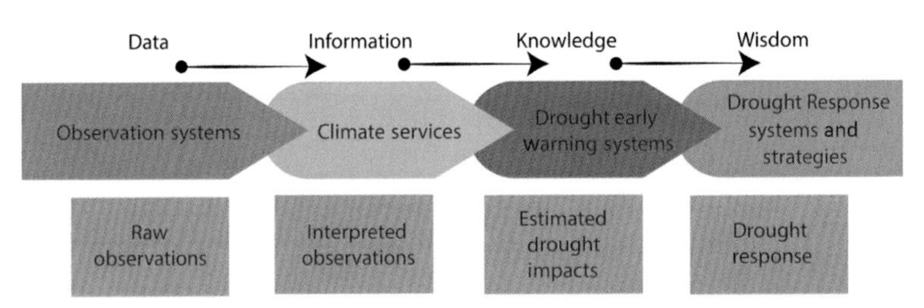

Figure 12.1 The data–information–knowledge–wisdom hierarchy as it might apply for effective DEWS. *DEWS*, Drought early warning system.

Drought Early Warning and Forecasting
DOI: https://doi.org/10.1016/B978-0-12-814011-6.00012-9

management responses. Information must be useful to be used, and it must be used to be useful.

In this context, raw observations (data) are translated into interpreted observations (information), which are then translated into estimated drought impacts (knowledge). These transformations also broadly align with observations systems, climate services, and DEWS. Each of these components builds upon the system below. One recurrent theme in this book has been the need to effectively combine multiple data and information sources to support estimates of drought impacts. The format of effective and actionable information often involves a translation into sector-specific impacts and recommendations. For example, observed and predicted meteorological conditions can be translated into specific potential forecasts of streamflow, reservoir levels, or snowpack. A similar transformation process can be used to predict agricultural outcomes or rangeland conditions. Such transformations can make drought information much more usable.

One excellent process for estimating drought impacts, discussed in Chapter 5, Tools of the trade 2—land surface models, involves running hydrologic and land surface models. These models can assimilate a broad suite of observations and translate this data into specific estimates of runoff, streamflow, or snowpack. This information can be used to guide decisions surrounding water use and storage. Another very common translation, discussed briefly in Chapter 9, Sources of drought early warning skill, staged prediction systems, and an example for Somalia, and Chapter 10, Practice—evaluating forecast skill, involves the translation of drought signals into estimated shocks to agricultural and pastoral/ranching livelihoods. Crop models or statistical relationships can often be used to translate environmental conditions into specific outcome-related information.

In all of these examples, abstract information is transformed into sector-specific information. For example, a 200-mm seasonal rainfall deficit is translated into a specific change in streamflow or crop yields. These latter drought impacts can be further interpreted as impacts associated with economic and health-related outcomes. Low streamflow values might reduce hydropower availability or increase the risk of acute diarrhea and cholera. Reduced yields might decrease farm incomes and increase food prices.

To be drought early warning system as opposed to drought warning system, systems must support early assessments of drought impacts. Early information arises from at least three sources. The first and easiest source is produced by the inherently lagged responses of many important drought impacts. Many regions of the world grow crops during the summer using snowmelt from winter storms. Rivers and streams generally experience low-flow conditions months after the peak period of precipitation. Almost all rainfed crops exhibit a fairly long delay between the critical periods of green vegetation growth and grain filling, and the time when the crops are actually harvested. The economic impacts of yield deficits, both for farmers and market prices, will also be delayed. Such lags mean that timely monitoring systems, in and of themselves, can provide effective early warning tools, especially when drought information is rapidly and accurately transformed into drought impact assessments.

Fig. 12.2 provides an example of such "easy" early warning, based on the maps of snow water equivalent for Afghanistan.[1] These maps are produced by the National Aeronautics and Space Administration (NASA) (https://ldas.gsfc.nasa.gov/fldas) using the Famine Early Warning Systems Network (FEWS NET) Land Data Assimilation System (FLDAS) (McNally et al., 2017). While setting up and running this instance of the NASA Land Information System is challenging, using it to anticipate drought impacts in Afghanistan can be quite straightforward, given the long lead time between Afghanistan's main period of precipitation accumulation (boreal winter) and crop production (boreal summer). In early 2018 NASA analysts[2] used the FLDAS to identify that a commonly used metric of snowpack—FLDAS-modeled estimates of February soil water equivalence—was the lowest on record since 2001. This information was used by FEWS NET to provide early warning.

The natural progression of the annual weather cycle, typically referred to as a region's "climate" or long-term average weather conditions, provides a second "easy" source of early warning skill. In general, almost all droughts evolve in interaction with the climatological sequence of precipitation. Except for a few regions of the tropics, most areas exhibit one or two distinct rainy or snowy seasons a year. As one approaches and passes the midpoint of these seasons, the type of integrated DEWS described in this book can provide increasingly accurate and precise sector-specific impact predictions. Quantifying the chance that streamflow, reservoir levels, snowpack, growing conditions, or fodder availability might recover from early or midseason water deficits can be relatively straightforward,

[1] https://nasaharvest.org/news/afghanistan-drought-research-highlights-harvest-partner-amy-mcnally

[2] <https://earthobservatory.nasa.gov/images/91851/record-low-snowpack-in-afghanistan>. Analysis performed by NASA scientists Amy McNally and Jossy Jacob.

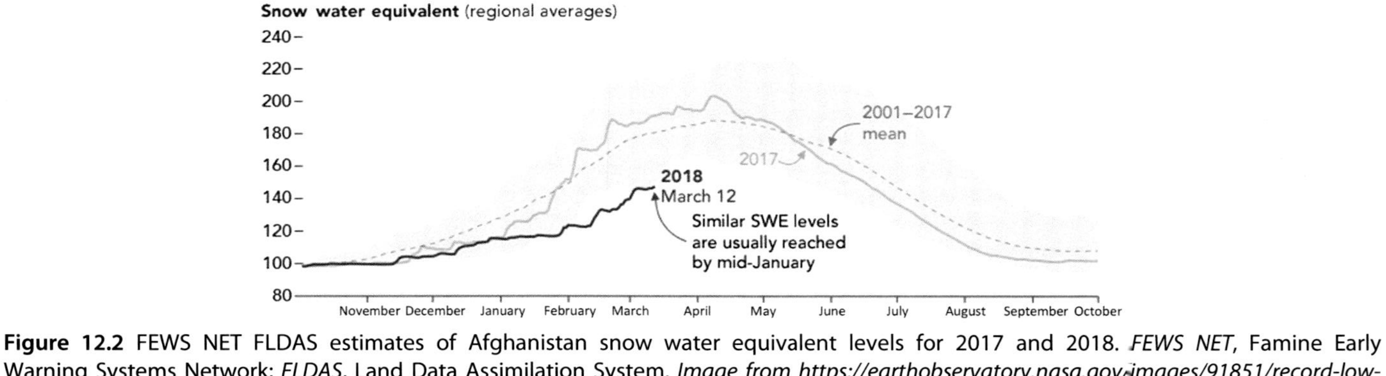

Figure 12.2 FEWS NET FLDAS estimates of Afghanistan snow water equivalent levels for 2017 and 2018. *FEWS NET*, Famine Early Warning Systems Network; *FLDAS*, Land Data Assimilation System. *Image from https://earthobservatory.nasa.gov/images/91851/record-low-snowpack-in-Afghanistan.*

yet also a very powerful source for early warning and effective response. This progression can be seen in Fig. 12.2. By early March the snow water equivalent for Afghanistan can be assessed through the middle of the season. Given that the value for 2018 was at an historic low, the chance of recovery was also very low.

Using historical data sets such as the Climate Prediction Center's African Rainfall Climatology version 2 (ARC2) product (Love et al., 2004), it is possible to assess the chance of a seasonal recovery by combining observed conditions with samples from past seasons. Such sampling typically assumes that all prior outcomes are equally probable, but other selection criteria might be used, such as selecting years with El Niño-Southern Oscillation conditions similar to those in a given season. Fig. 12.3, for example, shows seasonal ARC2 precipitation accumulations for October–December of 2019 from a region located in the maize triangle of South Africa. This is a very critical maize-growing region for southern Africa. The ARC2 rainfall observations extend to October 22. Seasonal accumulations beyond that point are simulated using all the available previous seasons (shown with thin *black* lines). While normal or above-normal October–December rains are possible for this region, the slow onset of rains appears likely to result in below-normal October–December precipitation totals.

Finally, as discussed in Chapter 9, Sources of drought early warning skill, staged prediction systems, and an example for Somalia, and Chapter 11, Practice—integrating observations and climate forecasts, climate and weather forecasts can be integrated with observations to peer into the future and help anticipate drought impacts. When these three sources of early warning skill are combined, seamless drought impact outlooks can be produced before, during, and after a season, supporting a defense-in-depth approach. In practice, in most places, the development of such systems remains a goal for the future. In theory, however, our current capacities should be able to support their development in the near future.

12.1 Actionable information and the three pillars

It should be realized, however, that effective DEWS need to be combined with the two other pillars of drought risk management: drought vulnerability and risk assessment, and drought preparedness, mitigation, and response (Fig. 12.4). Together with drought monitoring systems, these provide the key building blocks of an effective drought management

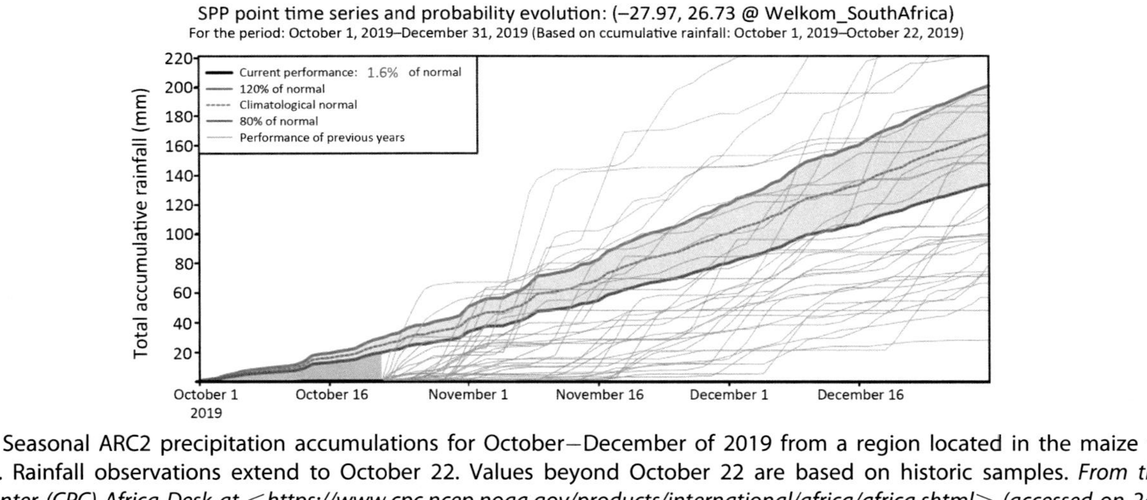

Figure 12.3 Seasonal ARC2 precipitation accumulations for October–December of 2019 from a region located in the maize triangle of South Africa. Rainfall observations extend to October 22. Values beyond October 22 are based on historic samples. *From the Climate Prediction Center (CPC) Africa Desk at <https://www.cpc.ncep.noaa.gov/products/international/africa/africa.shtml> (accessed on 24.10.19).*

Figure 12.4 The three pillars of drought risk management.

policy (Wilhite, 2011).[3] For example, Tadesse's (2016) white paper "Strategic Framework for Drought Risk Management and Enhancing Resilience in Africa" () provides an excellent description of how these components can fit together to inform effective policy. Effective drought management involves drought policy and governance for drought risk management; drought monitoring and early warning, drought vulnerability and impact assessment; drought mitigation, preparedness and response; knowledge management and drought awareness; and finally, reducing the underlying factors of drought risk.

Proactive drought risk management involves a thorough understanding of the local risks (Tadesse, 2016) associated with meteorological, agricultural, hydrologic, and socioeconomic droughts (Wilhite, 2000) (Fig. 12.5). Natural climate variability, potentially enhanced by human-induced climate change, can reduce precipitation or increase reference evapotranspiration (RefET) by increasing temperature, winds, incoming solar radiation, or reducing relative humidity. Changes in infiltration into the soil, runoff, interflow, and ground water recharge, combined with variations in actual evaporation and transpiration, can conspire to produce soil water deficiencies. These deficiencies, in turn, can lead to plant water stress, reduced biomass and yields, as well as reduced streamflows, and inflows to reservoirs, lakes, ponds, and wetlands. These anomalies can also negatively affect wildlife habitats.

Drought risk management seeks to understand and identify the various components of drought risk, along with the various strategies for managing these hazards (Hayes et al., 2004). Drought risk management can be broadly categorized into two fundamental categories of action: proactive risk reduction and reactive drought responses; it covers planning and response, managing both drought risks and impacts (Pulwarty and Sivakumar, 2014). Actionable drought information supports these objectives.

[3] http://www.droughtmanagement.info/pillars/

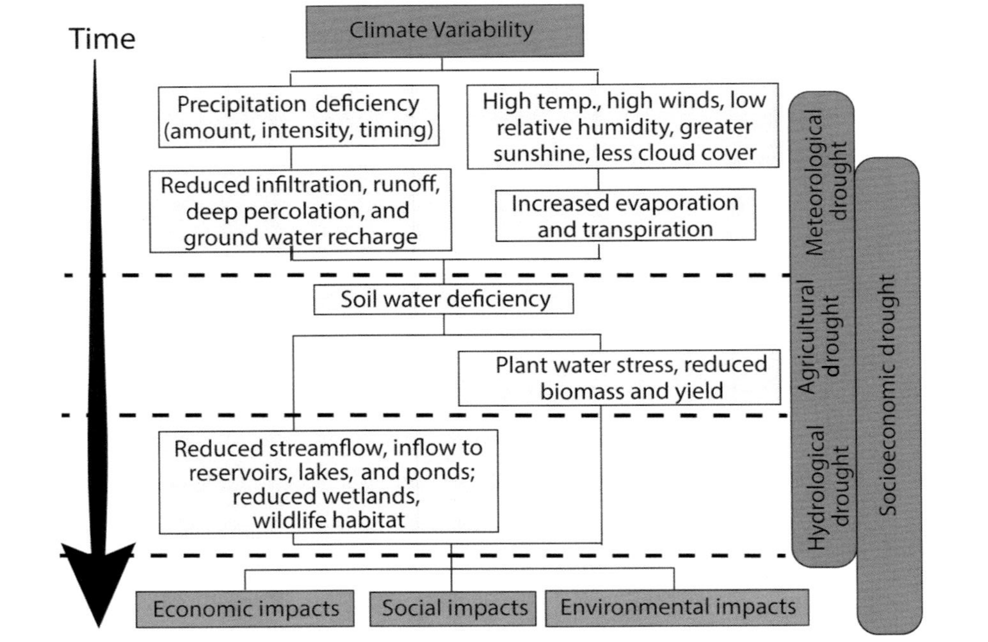

Figure 12.5 Relationships between meteorological, agricultural, hydrological, and socioeconomic droughts and their impacts. *After Figure 6 in Tadesse, T., 2016. Strategic framework for drought risk management and enhancing resilience in Africa. In: African Drought Conference based on a framework provided by the National Drought Mitigation Center, University of Nebraska-Lincoln, USA.*

12.2 Actionable information and decision-making networks—an example from famine early warning in East Africa

Effective DEWS provide information that supports real world responses. By providing early information, these systems can support actions that blunt negative drought impacts. So, the provision of actionable information often involves a translation into sector-specific impacts and recommendations. In practice, this almost always involves the supposition and interpenetration of multiple networks of human specialists. For example, moving from left to right across Fig. 12.1, we have Data → Information → Knowledge → Wisdom, or Observation systems → Climate services → Drought early warning systems → Drought response systems. On the left of this series, we find technical experts: satellite engineers, meteorologists running weather and climate models, and so on. Despite their technical focus, many members of this community will certainly desire that their products benefit society. This leads to the next layer, which contains specialists focusing the provision of climate services. Climate services provide environmental information and forecasts in accessible, timely formats suitable for societal applications. Experts within the next early warning layer translate this information into sector-specific impacts (Fig. 12.5). This book has emphasized the utility of a staged multiproduct approach to early warning. This staged approach provides multiple opportunities for drought detection. Just as important, however, is that this approach can provide the next set of experts—the final layer containing response systems and decision-makers—adequate time to put in place and execute a contingency plan. Even this "last mile" layer will typically contain both domain experts and policy and business decision-makers.

When they function correctly, these overlapping networks of human experts provide a tremendous example of our ability to work together collectively to guard lives and livelihoods. Dozens or even hundreds of experts may play a role in motivating an effective disaster response. Here, using the context provided by one specific multipartner activity, the FEWS NET (www.fews.net), we provide examples of actionable information in the context of the FEWS NET food security projection process. Created in 1985 by the U.S. Agency for International Development (USAID) after devastating famines in East and West Africa, FEWS NET provides objective, evidence-based analysis to help government decision-makers and relief agencies plan for and respond to humanitarian crises.

Analysts and specialists in 19 field offices work with U.S. Government science agencies, national government ministries, international agencies, and nongovernmental organizations (NGOs) to produce forward-looking reports on 28 of the world's most food-insecure countries. The "NET" in FEWS NET represents a vast network of internal and external partners, ranging from collaborators in data collection and analysis to consumers of FEWS NET reports. FEWS NET products include monthly reports and maps detailing current and projected food insecurity timely alerts on emerging or likely crises, specialized reports on weather and climate, markets and trade, agricultural production, livelihoods, nutrition, and food assistance.

The FEWS NET DEWS (Funk et al., 2019) is just one component of a much larger analytical framework that provides assessments of food-insecure populations. The DEWS supports monthly food security outlooks (Magadzire et al., 2017), which inform contingency planning and assessments of food insecurity (Verdin et al., 2005; Choularton, 2007; Brown, 2008). While drought monitoring and drought impact assessments play an important role in FEWS NET, they are just one potential driver of food insecurity. Adopting Aristotle's causal nomenclature (Fig. 12.6), we can describe the "ultimate" and "proximate" causes of severe food insecurity. Severe hunger can stem from multiple ultimate causes: in addition to drought, conflict, poverty, population pressure, poor governance, a lack of health services, and land degradation can lead to severe undernutrition. These and other factors almost always act in some combination, resulting in a hazardous combination of shock, exposure, and vulnerability (discussed in Chapter 6: Tools of the trade 4—mapping exposure and vulnerability). To cope with these complex interactions, FEWS NET food security analysts have developed a sophisticated food security scenario development process.[4] The ultimate goal of this process is to identify and quantify the number of extremely food-insecure people in the world's most food-insecure countries. To achieve this goal, FEWS NET analysts examine the proximate causes of food insecurity: food access, food availability, nutrition, and stability (Brown et al., 2015). These analysts form an international network that would be located at the right-hand side of Fig. 12.1.

Yet it is important to note that to the right of this network of analysts, an additional network of decision-makers who program and deliver humanitarian responses exists. This network comprises experts at the

[4] https://fews.net/sites/default/files/documents/reports/Guidance_Document_Scenario_Development_2018.pdf

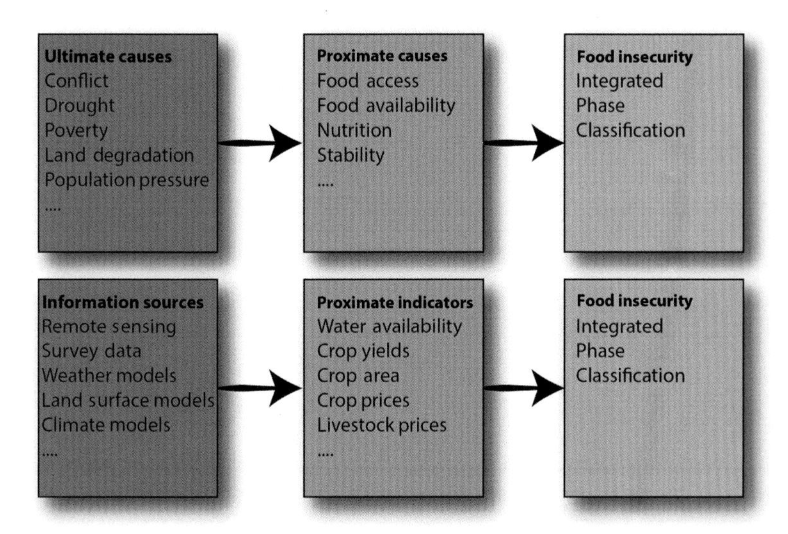

Figure 12.6 Ultimate and proximate causes of food insecurity, along with information sources and proximate indicators suitable for food.

USAID's Office of Food for Peace and the World Food Programme, for example. Such interconnecting layers of experts play a critical role in the effective use of the information provided by DEWS. Without food security analysts, even the most perfect drought information provided by the FEWS NET DEWS would be of limited value. These analysts consume DEWS information, evaluate the proximate drivers of food insecurity (food access, food availability, nutrition, and stability), and estimate specific levels of food insecurity, quantified using the tools and protocols of the Integrated Phase Classification (IPC) system (Frankenberger and Verduijn, 2011).

The FEWS NET food security scenario development process (see footnote 5) is based on a household (HH) level food security analysis. This procedure is based on six steps (Fig. 12.7). Step 1 identifies the specific geographic area of focus, the associated population, and the type of HH group to be analyzed. This group is chosen because it is likely to be very food insecure. Characterizations of the livelihoods of the target HH play a major role in the analysis. During Step 2, analysts characterize the current food security situation and describe HH level consumption and food access. What key assets do these HHs rely on to access food and income? What strategies do they use to gain income and access food? Finally, analysts use all the available information to assign the modeled

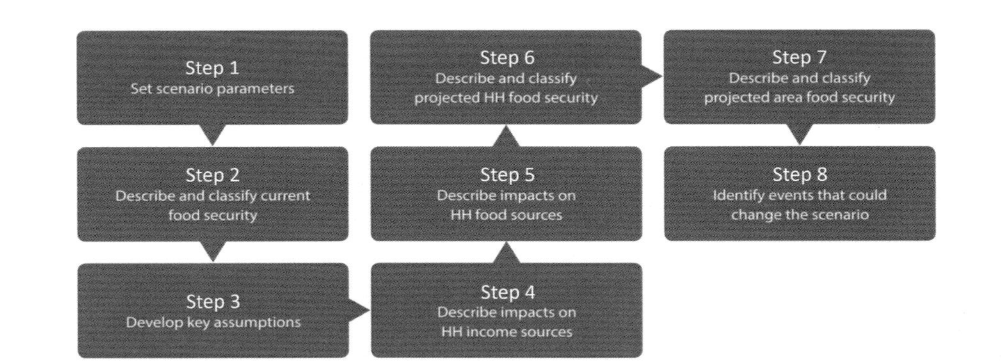

Figure 12.7 The FEWS NET food security scenario development process. *FEWS NET*, Famine Early Warning Systems Network. *From the FEWS NET guidance document <https://fews.net/sites/default/files/documents/reports/Guidance_Document_Scenario_Development_2018.pdf>.*

HHs an IPC score. A score of 2 indicates food stress. A value of 3 or 4 denotes a serious or very serious food security crisis or emergency. A value of 5 indicates famine conditions.

In the next (third) step, analysts provide key assumptions about what is likely to happen over the 4- to 8-month period. These assumptions often relate to rainfall performance, crop production, wage levels, water and pasture availability, livestock births, and staple food prices. Step 4 then relates these assumptions to HH incomes, which, along with food prices, strongly determine HH food access, that is, the ability to buy adequate and nutritious food. Step 5, in a similar fashion, relates the food security assumptions to impacts on HH food sources, which often strongly influence impacts on food availability, that is, the ability of a HH to directly procure food via crop, vegetable, and livestock production. The results of Step 4 and Step 5 are then used in Step 6 to make projections of the number of food-insecure people.

Fig. 12.8 shows an example of FEWS NET food-insecure population estimates for Ethiopia, Kenya, and Somalia, drawn from FEWS NET Food Assistance Outlook Briefs.[5] Following another poor March-to-May rainy season in 2019, these countries once again faced atypically high levels of acute food insecurity, based on numerous analyses by national and international food security analysts. This bar plot shows estimates of the total number of people anticipated to experience food insecurity, defined through classification at IPC Phase 3 or higher. FEWS NET projections are used for estimates beyond August 2019.

Two important features can be noted in this time series. *First*, the 2019 food security situation appears very concerning, with the size of the expected food-insecure population remaining similar to recent crises in 2016 and 2017. The 2016 and 2017 crises were associated with the strong 2015/16 El Niño and the 2016/17 La Niña, and related droughts (Funk et al., 2018). Despite only weak El Niño conditions, the 2019 March-to-May season was also very dry.[6] The recent dry conditions in the Horn of Africa have largely been observed in pastoral areas, where recovery is typically more protracted than in agricultural areas. While one good growing season may rapidly improve economic conditions for farmers, it can take years for herds of livestock to be built backup.

[5] http://fews.net/global/food-assistance-outlook-brief/

[6] http://blog.chg.ucsb.edu/?p = 592

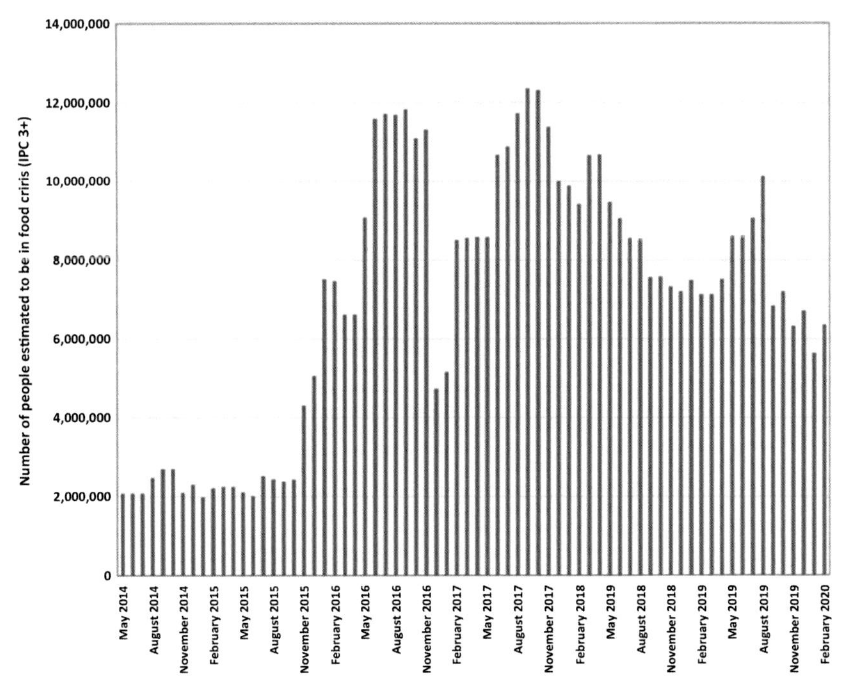

Figure 12.8 Consecutive FEWS NET Food Assistance Brief projections of food-insecure population totals for Ethiopia, Kenya, and Somalia, based on the average low and high estimates. *FEWS NET*, Famine Early Warning Systems Network.

Second, the data may indicate a large systematic increase in the frequency and magnitude of severe food insecurity. In 2014 and 2015 typical levels of acute food insecurity were at about 2 million people. Since 2015 this appears to have increased to around 6–10 million people. Following severe droughts in 2015, 2016/17, and 2019, FEWS NET has anticipated 10–12 million people in IPC Phase 3 or higher.

For Ethiopia and Kenya, these trends appear surprising, given estimates of annual incomes for the poorest 20% of the population (Fig. 6.9). These estimates are based on World Bank Development indicator statistics and have been produced by combining gross national income values, based on the Atlas method[7] and expressed in current U. S. dollars, population, and estimates of the income share held by the lowest 20%. These latter factors are only observed intermittently, and

[7] https://datahelpdesk.worldbank.org/knowledgebase/articles/77933-what-is-the-world-bank-atlas-method

linear interpolation has been used to produce continuous estimates. Ethiopia and Kenya, however, have recent surveys from 2015, during which the share of income estimates for Ethiopia and Kenya were just 6.6% and 6.2%—the poorest 20% in these countries only earn about 6% of the overall wealth. Considering these national scale indicators alone, one might anticipate a dramatic decline in vulnerability to food insecurity, since average incomes for poor HHs have increased by more than 300% since the early 1990s. These national averages, however, obscure important fluctuations in subnational HH-level incomes, as well as the potential implications of fluctuations in prices.

We can look at this same data in a different way by contrasting the average incomes in Kenya and Ethiopia with the incomes earned by the lowest 20% of the population (Fig. 6.10). What we see in this data is a substantial increase in the gap between middle class and poor populations. So, despite the fact that wealth is increasing, we are also seeing an increasing wealth disparity. These income gaps may be interacting with large fluctuations in commodity and food prices. Commodity prices in Ethiopia and Kenya remain very volatile, as illustrated by Fig. 6.11, which shows nominal wholesale maize prices. These values have not been adjusted for inflation and, hence, may overemphasize recent price increases. On the other hand, incomes for the poorest HHs may not be keeping pace with national inflation rates, as implied by Fig. 12.7. Ethiopian maize prices spiked in September–October of 2017 and have remained relatively high since. Kenya maize prices exhibit a clear periodicity that aligns with major recent droughts in 2010/11, 2016/17, and 2019. During these Kenyan price spikes, we see a doubling of maize prices. For poor HHs living on a dollar or two a day and spending 60%–70% of their HH income on food, such spikes can lead to dramatic food access limitations.

So, with this information as background, we next turn to an example focusing on agricultural production in Somalia, one of the most drought-sensitive countries in the world. Providing early and actionable quantitative predictions of potential drought impacts vis-à-vis regional annual cycles is one of the key objectives of modern DEWS. Using agricultural outcomes in Somalia as an example, we describe such a process, paying attention to three primary questions: (1) How can we use our preexisting knowledge to appropriately filter environmental data for the purposes of agricultural monitoring? (2) How can we translate environmental information into quantitative estimates of crop production? And finally, (3) When can we reliably provide such information?

Question 1 deals simultaneously with two important aspects of early warning—having a systematic approach to separating climate "noise" and climate "signal." In a region that is very wet, or unpopulated, or not planted, or out of season, anomalous environmental conditions may not relate at all to variations in crop yields, HH incomes, HH food availability, or variations in food prices. When assessing drought impacts, it is also important to consider drought exposure, which will be related to how, and if, the land and water from a given region are being used. It is common, therefore, to identify crop growing areas, such as those shown in the upper left of Fig. 12.9. What we see from this panel is that most of the crop production in Somalia occurs in just a small fraction of the country. If we are interested in deriving national estimates of crop production, focusing on this region will refine the accuracy of our results.

Next, we can use our a priori knowledge of crop water requirements to estimate the amount of water a healthy crop will require. Using assumptions from the Geospatial Water Requirement Satisfaction Index model[8] and RefET, we can estimate the ideal water requirement for the most common crop grown in Somalia (sorghum). Somalia has two growing seasons, a "Gu" season between March and May, and a "Deyr" season between October and December. Fig. 12.9 shows precomputed water requirements for each 10- or 11-day period in the second (October–December) rainy season. These values are derived by estimates of the crop stage and RefET. During the vegetative and grain filling phases of crop growth, the optimal water requirement matches the RefET. Before and after these stages the crop water requirement is less than the RefET value. A plant receiving this much water would assume to fully meet its water requirement.

While sorghum plants will draw this moisture from the soil, precipitation will likely provide a reasonable proxy indicator in rainfed regions. Simple comparisons (Fig. 12.9, right) of average Deyr rainfall conditions and crop water requirements can tell us a lot about typical Somali crop growing conditions. They are very poor, with average rainfall conditions only reaching about half the crop water requirements. From a food security perspective, this is important information. Most crop seasons in Somalia are likely to be poor. The season, furthermore, is typically very

[8] Support for this analysis was kindly provided by Will Turner, UC Santa Barbara Climate Hazards Center.

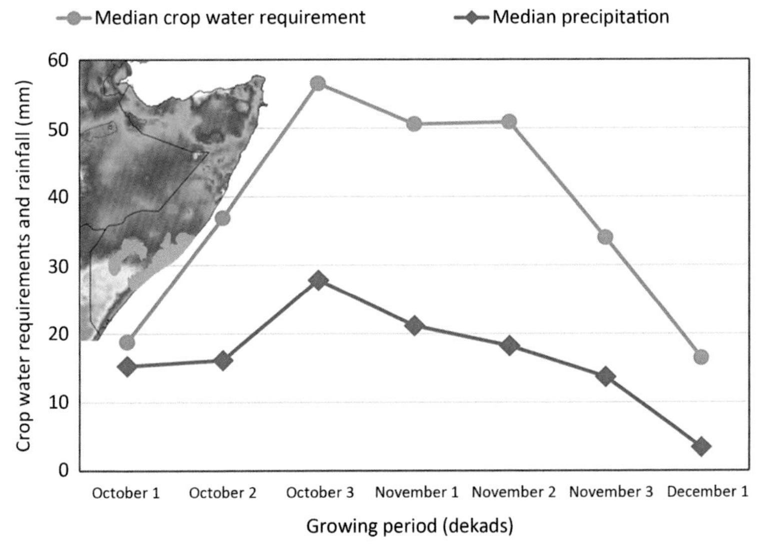

Figure 12.9 Typical (median) progression of sorghum crop water requirements and precipitation averaged over crop growing regions in southern Somalia. Crop growing zones are shaded in magenta and shown in the upper left. Background shading shows long-term average October–December rainfall.

short. This means that there is a very short window in which farmers can try to produce a healthy crop.

Under such conditions, we might expect to find strong relationships between moisture supplies and Somalia crop production. To test this hypothesis, we created a weighted sum of the observed CHIRPS2.0 precipitation, in the observed crop growing areas of Somalia. To create this sum, we used the same crop coefficients used to generate the water requirement estimates summarized in Fig. 12.9.

$$PR_{sum} = \sum_{i=1}^{7} k_i P_i$$

The crop coefficients for the seven 10- or 11-day periods (dekads) were 0.3, 0.65, 1, 1, 1, 0.65, and 0.3. The log-linear relationship between this weighted rainfall total and observed national 1995–2018 sorghum production was very strong (Fig. 12.10, left), with an R-squared value of over 0.75. One exceptionally wet season (1997) associated with flooding and low crop production outcomes was excluded. This data told a compelling story of the extremely broad range of agricultural outcomes. Good years, with 150 mm or more of rainfall, can produce more than

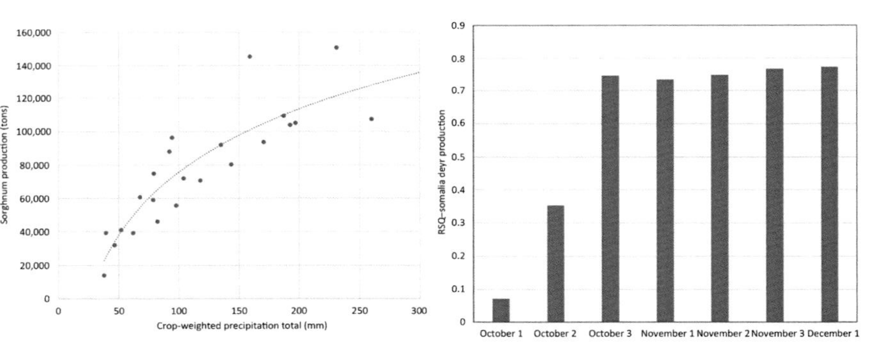

Figure 12.10 Left—logistic regression results relating crop-weighted seasonal precipitation totals and national Deyr sorghum production. Right—variance explained by successive models fits to totals starting on October 1 and ending on the dekads shown.

100,000 tons of sorghum. Poor years might produce only 40,000 or 20,000. The short growing period and large discrepancies between typical rainfall amounts and crop water requirements (Fig. 12.9) result in a very steep relationship between moisture supply and production values.

This relationship can clearly be used to assess crop production, given observed rainfall totals, producing more "actionable" information for FEWS NET analysts. Detailed and quantitative assumptions about crop production (Step 3 in Fig. 12.7) can directly inform projections related to HH-level food access and availability (Steps 4 and 5). Exploring the performance of our estimates over each dekad of the growing period (Fig. 12.10, right), we also note that very strong relationships appear exceptionally early in this region and season. By the first week of November, when we have all the data for October, our simple regression model performance saturates with R-squared values of over 0.75. Physically, this result makes sense. Crops require adequate time to mature, and the chance of this happening given poor October rains appears exceedingly low. This has important food security implications. Poor or failed crop seasons can be robustly identified in November. Note that our analysis of Normalized Difference Vegetation Index data (presented in Chapter 9: Sources of drought early warning skill, staged prediction systems, and an example for Somalia) produced very similar results.

12.3 Conclusion

Effective and actionable drought information systems are inherently composed of networks of partners. To support disaster risk reduction and drought risk management, experts from many different backgrounds and areas of expertise must work together effectively. This makes drought decision support challenging, but also very interesting. Diverse sets of decision-makers, ranging from large governmental, industry, and NGOs down to individual farmers, herders, and reservoir managers, all need effective drought information. This involves the development of networked teams of experts that run observation and modeling systems, provide climate services, and support DEWS. These systems can provide "actionable" information by translating environmental conditions into specific drought impacts (Fig. 12.5). Per capita water availability metrics and estimates of crop production deficits are two examples of these impacts. Other impact assessment models can be developed for different sectoral applications.

Even the best early warning systems and impact assessments are useless without strong links to drought response systems and strategies. This can involve the robust development of, and integration with, the two other pillars of drought risk management: drought vulnerability and risk assessment, and drought preparedness and mitigation (Fig. 12.4). Strong connections between technical DEWS partners and specialists designing and implementing disaster response and mitigation activities (Fig. 12.1) can ensure the robust application of drought early warning information. For food security applications, such as FEWS NET, the development, monitoring, and prediction of appropriate proximate indicators of food access and food availability can make food security projections more accurate and timely. An example focused on Somalia showed how an agricultural modeling lens can be used to filter environmental data in space and time, producing both insights into Somalia's severe food insecurity and highly accurate estimates of national crop production.

References

Brown, M.E., 2008. Famine Early Warning Systems and Remote Sensing Data. Springer Science & Business Media.

Brown, M., Antle, J., Backlund, P., Carr, E., Easterling, W., Walsh, W., et al., 2015. Climate Change, Global Food Security, and the U.S. Food System. USDA, p. 146.

Choularton, R., 2007. Contingency Planning and Humanitarian Action: A Review of Practice, Humanitarian Practice Network.

Frankenberger, T.R., Verduijn, R., 2011. Integrated Food Security Phase Classification (IPC).

Funk, C., Shukla, S., Thiaw, W.M., Rowland, J., Hoell, A., Mcnally, A., et al., 2019. Recognizing the Famine Early Warning Systems Network: Over 30 Years of Drought Early Warning Science Advances and Partnerships Promoting Global Food Security. Bulletin of the American Meteorological Society, 100 (6), 1011–1027. https://doi.org/10.1175/bams-d-17-0233.1.

Funk, C., Harrison, L., Shukla, S., Pomposi, C., Galu, G., Korecha, D., et al., 2018. Examining the role of unusually warm Indo-Pacific sea surface temperatures in recent African droughts. Q. J. R. Meteorol. Soc. 144, 360–383.

Hayes, M.J., Wilhelmi, O.V., Knutson, C.L., 2004. Reducing drought risk: bridging theory and practice. Nat. Hazards Rev. 5 (2), 106–113.

Love, T.B., Kumar, V., Xie, P., Thiaw, W., 2004. A 20-year daily Africa precipitation climatology using satellite and gauge data. In: Proceedings of the 84th AMS Annual Meeting, vol. Conference on Applied Climatology, Seattle, WA–(CD-ROM).

Magadzire, T., Galu, G., Verdin, J., 2017. How Climate Forecasts Strengthen Food Security. World Meteorological Organisation Bulletin – Special Issue on Water 67.

McNally, A., Arsenault, K., Kumar, S., Shukla, S., Peterson, P., Wang, S., et al., 2017. A Land Data Assimilation System for sub-Saharan Africa food and water security applications. Sci. Data 4, 170012.

Pulwarty, R.S., Sivakumar, M.V., 2014. Information systems in a changing climate: Early warnings and drought risk management. Weather Clim. Extremes 3, 14–21.

Tadesse, T., 2016. Strategic framework for drought risk management and enhancing resilience in Africa. In: African Drought Conference.
Verdin, J., Funk, C., Senay, G., Choularton, R., 2005. Climate science and famine early warning. Philos. Trans. R. Soc. B: Biol. Sci. 360 (1463), 2155–2168.
Wilhite, D.A., 2000. Preparing for Drought: A Methodology.
Wilhite, D.A., 2011. Essential Elements of National Drought Policy: Moving Toward Creating Drought Policy Guidelines.

CHAPTER 13

Final thoughts

Some 3.2 million years ago, our common ancestor Lucy walked semierect across eastern Africa (Fig. 1.1). Around 100,000 years before the present, the range of *Homo sapiens* was still limited to southern and eastern Africa (Fig. 1.2). In the next 30 years, a mere second on the Earth's geological clock, our world population is expected to reach nearly 10 billion people, but our warming planet may likely struggle to support twice the food production of today to meet the rising demand (Ray et al., 2013). This expansion in population will expose — is exposing — hundreds of millions of people to increased risks and spikes in disaster frequencies such as droughts (Fig. 6.2). Managing these crises will require that we improve all three pillars of drought management systems (Fig. 12.4): drought early warning systems (DEWS); drought vulnerability and risk assessment; and drought preparedness, mitigation, and response (Wilhite, 2011).[1] Given that our changing climate is bringing increases in drought frequency, severity, and duration that impact an ever-increasing number of sectors, now is the time for improved drought risk reduction (Wilhite et al., 2014). Responding to this spiral will require a proactive approach that actively manages and mitigates droughts and drought vulnerability through effective national policies (Wilhite et al., 2014). Droughts do not necessarily have to be emergencies (Wilhite et al., 2014), but passively reacting to drought impacts simply treats the symptoms, rather than root causes, locking in an "hydro-illogical cycle." Such a cycle involves a period of crisis and disaster, rapid, expensive, and relatively unsuccessful response, followed by a period of forgetting and inaction.

Among Earth's inhabitants, humans are unique in their capacity to plan ahead, though we do not always do so. Because droughts are complex multiscale hazards, planning for drought requires complex layers of planning both coordination and response (Pulwarty and Sivakumar, 2014): "To cross the spectrum of potential drivers and impacts, drought information systems have multiple sub-systems which include an integrated risk assessment, communication and decision support system of

[1] http://www.droughtmanagement.info/pillars/

Drought Early Warning and Forecasting
https://doi.org/10.1016/B978-0-12-814011-6.00013-0

which early warning is a central component and output." Early warning systems, however, are much more than just forecasts, they must be linked to risk assessment, communication, and decision support systems (Pulwarty and Sivakumar, 2014). In this book, we have attempted to bring together many of the important pieces of DEWS—describing in one location the basic tools of the trade: observations, forecasts, climate, and land surface models; exposure and vulnerability assessments, drought indices; and the impacts of warming—and placed them alongside some practical applications.

While essentially doomed to be incomplete, given the scope of early warning systems, our goal has been to provide an accessible "one-stop-shop" entry point for practitioners or potential practitioners. Because effective DEWS requires overlapping communities of practice that span the data–information–knowledge–wisdom hierarchy (Fig. 12.1), it is useful to bring together discussions involving observation systems, climate services, DEWS, and drought response systems. Data providers on the left of Fig. 12.1 can benefit from understanding how DEWS works and what properties are required to support drought early warning. Modest systematic nonstationary errors, for example (as discussed in Chapter 8: Theory—Indices for Measuring Drought Severity), can wreak havoc in DEWS, leading to false alarms or missed opportunities for drought forecasts or detection. The seasonal averaging used to identify droughts can magnify systematic errors, and homogeneous, stable monitoring and forecast systems are therefore very important technical inputs for DEWS.

Response planners and responders on the other right-hand "wisdom" side of the spectrum shown in Fig. 12.1 can benefit from a better understanding of how our Earth system provides us opportunities for monitoring and prediction, and how models and observations can be leveraged to capitalize on these chances to make a difference. This perspective, for example, highlights the first half of precipitation seasons as a key period of maximum potential forecast skill. At mid-season, antecedent precipitation, temperature, radiation, and soil moisture conditions strongly influence the land surface, snowpack, and vegetation conditions. Two-week weather forecasts, which are typically quite accurate, give us an outlook for the heart of the rainy season, and climate forecasts can help use peer a little further ahead into the next several months. These sources of predictive skill can be rendered "interoperable" via appropriate downscaling techniques (Chapter 11: Practice—Integrating Observations and Climate

Forecasts) and used to drive impact models, such as land data assimilation systems (Chapter 5: Tools of the Trade 2—Land Surface Models).

Because planning for drought requires complex layers of planning both coordination and response (Pulwarty and Sivakumar, 2014), the technical components of 21st century's DEWS need to mimic the complex multiscale nature of droughts themselves, operating on multiple time frames and spatial scales. The multiscale drivers of drought demand multiscale early warning systems, and to this end, we have advocated a staged approach to DEWS (Fig. 3.7). Such a system includes drought risk assessment prior to the onset of precipitation season, followed by long-lead climate-based drought outlooks, mid-season assessments, and postseason evaluations (Fig. 9.6). Such integrated approaches draw on a wonderfully broad suite of available tools for monitoring, predicting, and assessing the many varied impacts of meteorological, agricultural, hydrological, and socioeconomic droughts (Wilhite, 2000) (Fig. 12.5). Like a surgeon, expert drought analysts will use a broad variety of tools in different drought stages and contexts. This breadth has an additional value when viewed from a signal detection perspective (Fig. 9.1). Multiple sources of information provide multiple opportunities to identify emergent drought crises, greatly reducing the chance of failure. A well-structured integrated DEWS should always identify large droughts; the question should be only "when" not "if." The Earth system is chaotic, and long-lead climate-based outlooks will often fail. Late-season recoveries or faulty monitoring data may confound mid-season assessments. But by the end of a season, a convergence of evidence approach bolstered by the power of 21st century satellite-based observing systems should catch almost all extreme events.

Two added benefits of a staged multi-source, multi-scale approach to developing 21st century DEWS are that (1) this approach can benefit from the predictive skill arising from multiple sources — slowly varying oceans, persistent land surface, soil water and vegetation anomalies, and weather-scale circulation anomalies, while also (2) capturing the potential climate change impacts associated with these systems as our oceans, land, and atmosphere warm.

While the science, resources, and techniques are fairly well developed to support such systems, more needs to be done to develop such capacity in developing nations (Tadesse, 2016). There is a great need, but also a great opportunity, to improve links between national and global early warning systems. To function well, DEWS can really benefit greatly by taking a network-based approach, drawing information sources from all across the internet. When the anesthesiologist puts us under for an important operation, we are not really concerned about whether the surgeon's scalpel was made in our country. By analogy an effective DEWS can draw from a myriad of sources of information to provide effective decision support.

To help guide such efforts, this book has focused on integrating observations with forecasts to provide actionable impact assessments. Building on our work with one global system (Famine Early Warning Systems Network), we have tried to describe how resources provided by global systems (satellite rainfall estimates, vegetation observations, weather and climate forecasts, and land surface model simulations) can be translated into localized sector-specific indices and impact models (Chapter 8: Theory—Indices for Measuring Drought Severity, and Chapter 9: Sources of Drought Early Warning Skill, Staged Prediction Systems, and an Example for Somalia). During the middle of a growing season, all the sources of predictive skill (ocean, atmosphere, land) contribute, and our ability to foresee drought impacts often reaches a peak trade-off between certainty and utility. Capturing that information often requires downscaling weather and climate forecasts and combining them with fine-resolution observations (Chapter 11: Practice—Integrating Observations and Climate Forecasts). With the advent of the internet, the data and modeling resources of many global early warning systems can now be tailored to serve the needs of many national and even subnational early warning systems.

DEWS, when they successfully link across sectors and communities can empower a relatively small number of people to make critical decisions that positively impact hundreds of thousands of people and animals, or vast regions of forest, farm, or rangeland. The rapid advance of computing, satellite observation, and modeling systems makes such decision support increasingly possible. To achieve this goal, however, the early warning practitioners must be spread along a continuum that effectively links data with decisions and information with wise actions. Drought early warning is ultimately about interpreting our world and speaking to each other in ways that allow us to make sense of the future, enabling actions that reduce both future impacts and future risks. Effective DEWS involves our relationships with time, but also with each other. None of us can do this alone, but together, we can always do it better. Continued advances in 21st century drought early warning science will be welcomed by a thirsty and increasingly warm world.

References

Pulwarty, R.S., Sivakumar, M.V., 2014. Information systems in a changing climate: early warnings and drought risk management. Weather Clim. Extremes 3, 14–21.

Ray, D.K., Mueller, N.D., West, P.C., Foley, J.A., 2013. Yield trends are insufficient to double global crop production by 2050. PLoS One 8 (6), e66428. Available from: https://doi.org/10.1371/journal.pone.0066428.

Tadesse, T., 2016. Strategic framework for drought risk management and enhancing resilience in Africa. In: African Drought Conference.

Wilhite, Donald A., "Chapter 35 Preparing for Drought: A Methodology" (2000). Drought Mitigation Center Faculty Publications. 72. http://digitalcommons.unl.edu/droughtfacpub/72

Wilhite, D.A., 2011. Essential elements of national drought policy: moving toward creating drought policy guidelines.

Wilhite, D.A., Sivakumar, M.V.K., Pulwarty, R., 2014. Managing drought risk in a changing climate: The role of national drought policy. Weather Clim. Extremes 3, 4–13.

Index

Note: Page numbers followed by "*f*" and "*t*" refer to figures and tables, respectively.

I

J

L

M

N

S

T

U

Printed in the United States
By Bookmasters